MATERIALS AND PROCESS SELECTION IN ENGINEERING

MATERIALS AND PROCESS SELECTION IN ENGINEERING

MAHMOUD M. FARAG

B.Sc., M. Met., Ph.D. (Sheffield)

Professor of Materials Engineering,
American University in Cairo, Egypt

APPLIED SCIENCE PUBLISHERS LTD
LONDON

APPLIED SCIENCE PUBLISHERS LTD
RIPPLE ROAD, BARKING, ESSEX, ENGLAND

British Library Cataloguing in Publication Data

Farag, Mahmoud M
Materials and process selection in engineering.
1. Materials
I. Title
620.1′1 TA403

ISBN 0-85334-824-3

WITH 96 TABLES AND 34 ILLUSTRATIONS

Printed in Great Britain by Galliard (Printers) Ltd, Great Yarmouth

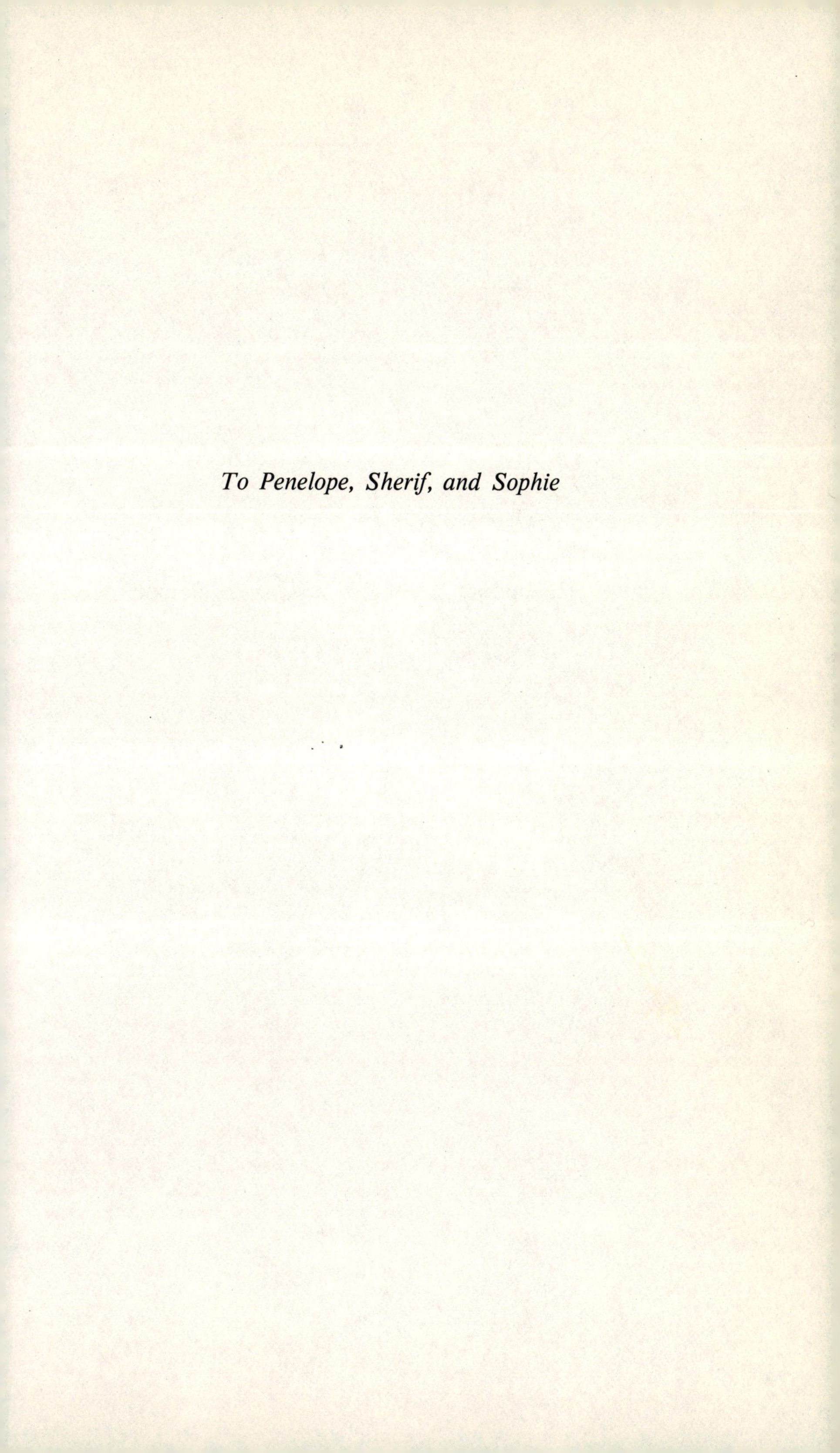

To Penelope, Sherif, and Sophie

Preface

One of the most important requisites for the development of a satisfactory product at a competitive cost is the selection of the optimum material from the multitude of available engineering materials. This task is not an easy one and a large proportion of product failures in service are due to incorrect use of materials. The task of selection is further complicated by the fact that the processes used in manufacturing the product have considerable influence on the properties of the materials and their behaviour in service. Consequently, materials should not be selected in isolation from the manufacturing processes, which, in turn, should not be selected in isolation from the product design and development processes.

The material and process selection activity should involve not only the technical aspects but also the economic aspects. This is because an essential prerequisite of a successful product is economic feasibility.

The objective of this book is to provide both the technical and economic backgrounds that will enable the engineer to select the optimum material–process combination for a given application. Therefore, the subject matter is presented from the point of view of the material selector. Information that is of a practical nature and quantitative methods that will help the reader in selection are provided at the expense of details that are of academic interest only. The first five chapters discuss the characteristics and processing of metallic alloys, polymers, ceramics, materials of construction, and composite materials. The effect of service conditions on these materials and a detailed discussion of their reliability in service are given in Chapters 6 and 7. In Chapters 8 to 10 the economic parameters that affect the cost of materials and processing are discussed. The technical and economic principles are then combined in Chapters 11 to 13, to establish the general principles and procedure for selecting materials and processes for engineering applications. The different principles that are established in the first thirteen chapters are applied to a variety of case studies in Chapters 14 to 20.

The discussions in the book are written at the level of post-graduate or final-year undergraduate engineering students, and practising engineers will also find the subject matter interesting and useful. The SI units are used throughout this book, but seven appendices are provided to give easy conversion to Imperial units. As much information as possible concerning the properties of the different materials is included, usually in tabular form for easy reference. Whenever possible the cost of materials is included as relative values rather than actual $ or £ per unit weight or volume, which date quickly. However, the reader will need to have access to other sources of data such as those described in Section 12.12.

The task of writing this book has been mainly one of collecting the relevant facts and information from the literature and the main sources are listed at the end of each chapter.

I owe much to my wife for her patience and help during the preparation of the manuscript.

MAHMOUD M. FARAG

Contents

1

Metallic Materials

1.1 INTRODUCTION

Pure metallic elements have a wide range of properties. Their alloys are even more versatile and form a large proportion of engineering materials. The major characteristics of metallic materials are their crystallinity, conductivity to heat and electricity, and relatively high strength and toughness.

Metallic materials can be divided into ferrous and non-ferrous alloys. Ferrous alloys have iron as the base metal and range from plain carbon steels, containing more than 98% iron, to high alloys containing up to about 50% of a variety of alloying elements. All other metallic materials fall into the non-ferrous category, which can be subdivided into light alloys, e.g. aluminium, magnesium, titanium and beryllium; heavy alloys, e.g. copper, zinc, lead and tin; refractory metals, e.g. tungsten, tantalum and molybdenum; and precious metals, e.g. gold, silver and platinum.

Metallic materials can be divided according to the method of production into cast alloys and wrought alloys. Cast alloys form about 20% of all industrial metallic materials and are cast directly into shape. Wrought alloys are usually shaped by hot or cold working into semi-finished products like plates, sheets, rods, wires or tubes. These semi-finished materials provide a starting point for the fabrication of finished components by further forming or machining processes. A third type, which has gained industrial favour in recent years, is powder metals and alloys. The powders are compacted and sintered to produce ready-to-use components that need very little further machining.

In most engineering applications, selection of the optimum metallic material is usually based on more than one of the following considerations:

1. Mechanical properties, as ordinarily revealed by the tensile, fatigue, hardness and impact tests.
2. Physical and chemical properties, such as specific gravity, thermal and electrical conductivities, thermal expansion coefficient, and corrosion resistance.
3. Metallurgical considerations, such as anisotropy of properties, hardenability of steels, grain size, and consistency of properties, i.e. absence of segregations and inclusions.
4. Processing considerations, such as formability, machinability and weldability.
5. Cost and availability.
6. Sales appeal such as colour and lustre.

The first four considerations will be discussed in this chapter and the fifth and sixth in Chapters 8, 9 and 10.

1.2 DEFORMATION OF METALLIC MATERIALS

Within the elastic range, the strain caused by external stresses is a result of changes in the dimensions of the material unit cell. The elastic modulus, the ratio between the stress and strain, is a characteristic of the material and is not markedly affected by alloying. All elastic moduli decrease with increase in temperature. The elastic modulus of a multiphase or a composite material is a function of the elastic moduli of the constituents, their amounts and arrangement, as will be discussed in Chapter 5.

After reaching the elastic limit, metallic materials usually continue to deform plastically by slip, and this involves dislocation movement. The strength of a material in the plastic range is determined by the forces required to move the dislocations. Therefore any hindrance to dislocation movement will increase the strength of the material. Distortions in the lattice, caused by foreign atoms in solid solution, strain-hardening, grain boundaries, or second phase precipitates, can act as obstacles to dislocations and are commonly used to enhance the mechanical properties of metallic materials, as will be discussed in Section 1.3.

1.3 STRENGTHENING OF METALLIC MATERIALS

The useful strength of a material is usually limited either by plastic yielding for ductile materials or by fracture for brittle materials. For many engineering alloys, an increase in strength is accompanied by a reduction in ductility, and the optimum combination of these properties should be selected to meet the service requirements. Most of the industrially used methods of strengthening are based on hindering the movement of dislocations in the material as outlined in Section 1.2.

Strengthening by forming a solid solution is commonly used in practice, especially for aluminium and copper alloys, e.g. Al–Mg and Cu–Zn alloys. Figure 1.1 illustrates the effect of solid solution alloying on the mechanical properties of Al–Mg alloys. The effectiveness of an alloying element in

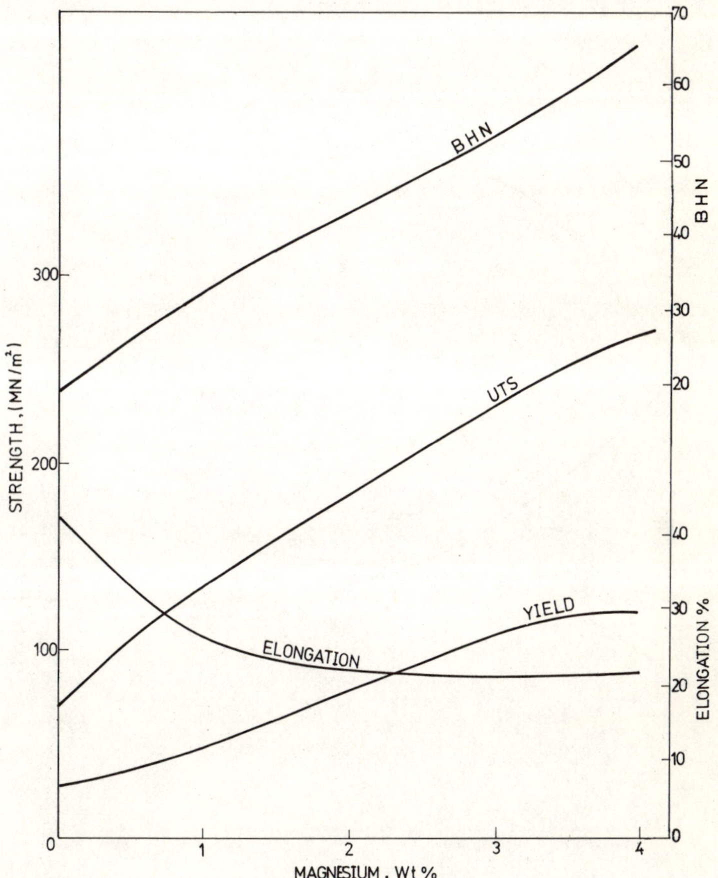

FIG. 1.1. Effect of magnesium addition on the mechanical behaviour of aluminium (solid solution hardening).

solution hardening is a function of the lattice distortion caused by atomic size mismatch between the parent metal and the solute.

Strain-hardening by cold working of wrought alloys is another important method of strengthening engineering alloys. Cold working increases the number of dislocations, causing them to tangle up and become more difficult to move. Figure 1.2 shows the effect of cold working on the mechanical behaviour of a plain carbon steel, AISI–ASA 1040. The rate of strain-hardening is affected by the lattice structure, grain size, impurity atoms and the phases present in the alloy. Severely strained materials usually exhibit

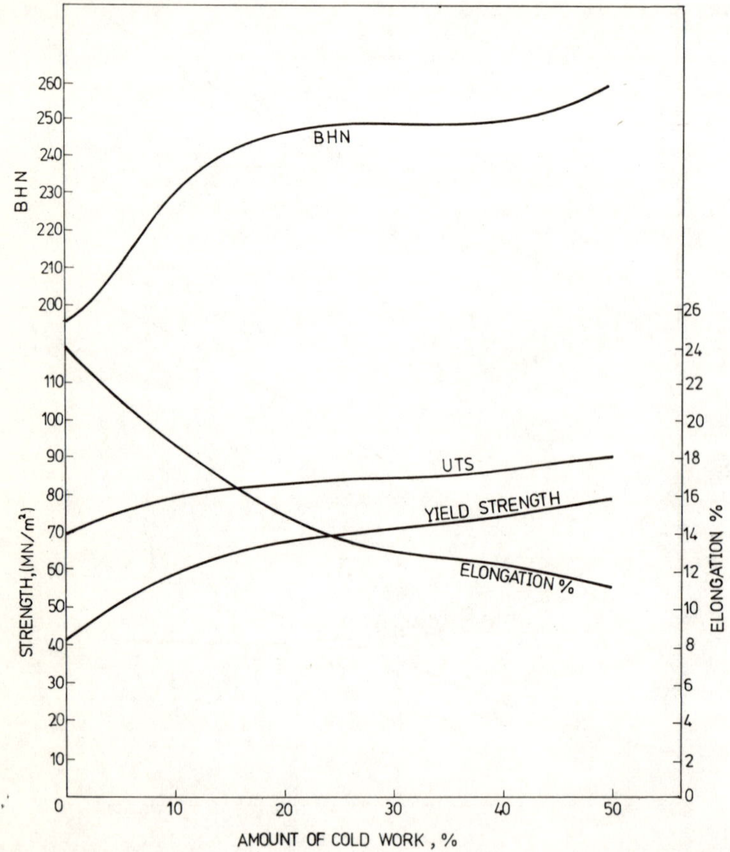

FIG. 1.2. Effect of reduction by cold working on the mechanical behaviour of AISI-ASA 1040 steel (0·4% C).

some anisotropy in mechanical behaviour that can be advantageous in some applications. Annealing the material by heating to above the recrystallisation temperature eliminates the effect of strain-hardening and restores the material to the soft condition.

The strength of metallic materials can usually be increased by reducing grain size. The quantitative relation between yield strength (Y) and grain diameter (d) is given by the Hall–Petch relationship:

$$Y = \mathrm{A} + \mathrm{B}\ d^{-1/2} \tag{1.1}$$

where A and B are constants, usually determined experimentally for a given material. In ultra-fine-grained steels, special treatments produce grain sizes of about 1 μm and increase the strength by approximately a factor of two.

A direct way of retarding dislocation movement is to introduce a hard, dispersed phase in the matrix. The dispersed phase is usually introduced by precipitation from a supersaturated solution. A basic requirement for any precipitation hardening alloy is a sloping solvus line (solid solubility curve) in the phase diagram, and one of the best known examples is the Al–4·5 Cu alloy. The precipitation hardening process involves a solution treatment and quenching—to produce a supersaturated solid solution—followed by ageing to precipitate the hard phase under controlled conditions. Combinations of precipitation hardening and strain-hardening can give more strengthening than either process alone.

The above analysis shows that specifying the chemical composition of an alloy is not enough to define its mechanical properties; the way it is treated (temper) should be also specified. An example is given in Table 1.1 of the temper designations for aluminium alloys.

In multiphase materials, an overall property depends on the physical interaction of the individual phases. When such materials are loaded the stresses within each phase depend on its distribution and shape, as well as on its relative mechanical properties with respect to surrounding phases. The behaviour of multiphase materials will be discussed in more detail in Chapter 5. Many steels can be classified as multiphase materials, and in the simple case of annealed plain carbon steels the hard iron carbides strengthen the softer ferrite matrix, as shown in Fig. 1.3. Increasing the carbide content increases the hardness, the yield strength and the ultimate tensile strength, at the expense of ductility and toughness. Changing the carbide shape from lamellar to spheroidal reduces the strength but improves the toughness of a given steel.

Strengthening of metallic materials can also be accomplished by replacing softer phases by harder ones through heat treatment, as in the case of steels,

TABLE 1.1

TEMPER DESIGNATIONS FOR ALUMINIUM ALLOYS

Temper designation	*Condition*
F	As fabricated, e.g. as rolled.
O	Annealed, recrystallised, softest temper.
H	Strain-hardened.
H1	Strain-hardened by cold working. A second digit indicates the degree of hardness, e.g. H12, quarter hard; H14, half hard; H18, full hard; H19, extra hard.
H2	Strain-hardened and partially annealed. H24, half hard; H28, full hard.
H3	Strain-hardened and stabilised by a low temperature thermal treatment.
T	Heat treated.
T2	Annealed castings.
T3	Solution treated followed by strain-hardening and natural ageing.
T4	Solution treated followed by natural ageing.
T5	Artificially aged with no solution treatment.
T6	Solution treatment followed by artificial ageing.
T8	Solution treated followed by strain-hardening and then artificial ageing.
T9	Solution treated followed by artificial ageing and then strain-hardened.

where a large variety of microstructures result from variations in the cooling rate from the austenitic region (γ-phase). Table 1.2 and Fig. 1.4 summarise the different heat treatment possibilities and the resulting microstructures and properties. The shape and position of the transformation curves (T_s, T_f) are functions of the chemical composition of steel. The critical cooling rate is defined as the slowest cooling rate that avoids the knee of the T_s curve and can be used as an estimate of the hardenability of a given steel. The lower the critical cooling rate, the higher the hardenability, and the easier it is to achieve martensitic structure in a given steel.

1.4 FERROUS ALLOYS

As indicated earlier, metallic materials can be grouped into ferrous and non-ferrous alloys. Ferrous materials, of which steels are the major alloys, form

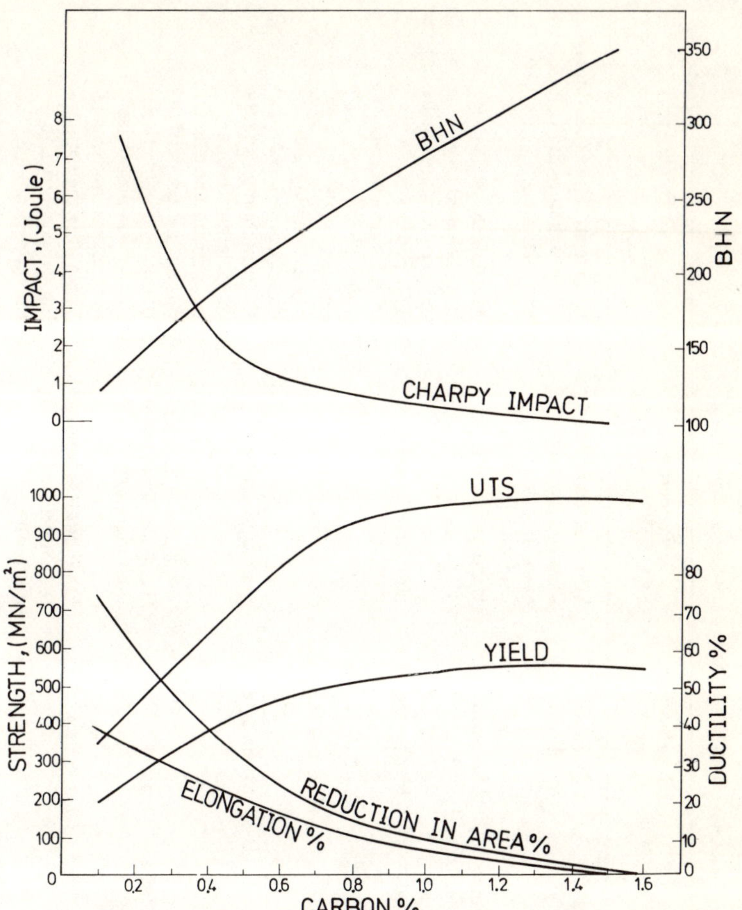

FIG. 1.3. Effect of carbon content on the mechanical behaviour of hot rolled 25 mm diameter bars of plain carbon steel.

about 90% of the total usage of metallic materials in the world. The main reasons for this overwhelming preference for ferrous materials are their versatile properties and their low cost.

Plain carbon steels are the cheapest and most commonly used steels and contain up to about 1·0% C, 1·65% Mn, 0·6% Si, 0·05% S and 0·04% P. Different countries adopt different identification systems and an example is shown in Table 1.3 of the AISI and SAE designations for plain carbon and low alloy steels. The strength of plain carbon steels is primarily a function of

TABLE 1.2

TRANSFORMATIONS AND PROPERTIES OF STEEL

Process	*Procedure*	*Resulting phases*	*Properties*	*Remarks*
Annealing	Slow cool from γ	$\alpha + Fe_3C$	Soft, ductile and tough	Coarse structure (pearlite)
Normalising	Air cool from γ	$\alpha + Fe_3C$	Relatively stronger than annealed steel with lower ductility	Finer structure than annealed steel
Quenching	Rapid cool from γ	Martensite	Hard and brittle	Martensitic steels must be toughened by a tempering treatment before use
Interrupted quench	Quench followed by slow cool through M_s and M_f	Martensite	Hard and brittle	This process is called martempering and is less severe and less likely to cause cracking than direct quench to M_s
Ausforming	Deformation just above M_s then slow cool	Martensite	Hard and brittle	Deformation during the quench interruption adds strain-hardening to martensitic properties
Austempering	Interrupted quench then isothermal transformation	$\alpha + Fe_3C$	Hard but less brittle than martensite	Bainite is obtained and has similar properties to tempered martensite
Tempering	Reheating of martensite	$\alpha + Fe_3C$	Hard but less brittle than untreated martensite	The higher the tempering temperature and the more prolonged the tempering time, the less the hardness and the greater the toughness

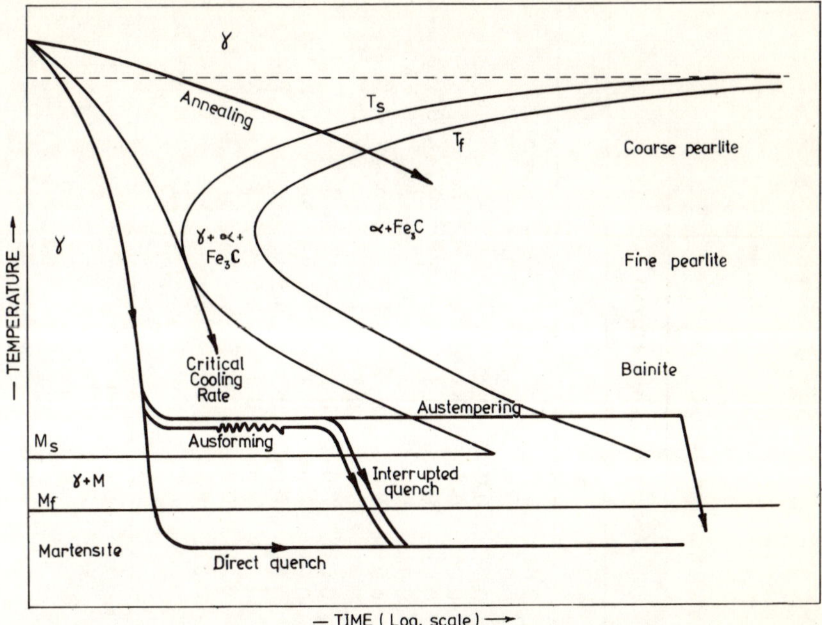

FIG. 1.4. Temperature-time-transformation diagram illustrating the effect of the mode of cooling on the resulting microstructure of steels. Interrupted quench results in martensitic structure, but with less internal stresses than direct quench. Annealing results in pearlitic structure and austempering in bainitic structure. T_s = transformation starts, T_f = transformation is completed, M_s = martensitic transformation starts, M_f = martensitic transformation is completed.

their carbon content, as shown in Fig. 1.3. Unfortunately, the ductility of these steels decreases as the carbon content is increased, and their hardenability is relatively low. Also, plain carbon steels lose most of their strength at high temperatures, become too brittle at low temperatures, and are subject to corrosion in most environments. These limitations can be eliminated by adding the appropriate alloying elements.

Steels that do not contain more than a 5% total of combined alloying elements are called low alloy steels and are designated as shown in Table 1.3. The commonly used alloying elements and their effect on the steel properties are given in Table 1.4. Since there are more elements to be kept within specified limits in alloy steels, they require better quality control and consequently are more expensive than plain carbon steels. However, alloy steels generally have better properties than plain carbon steels, as shown in Table 1.5.

TABLE 1.3

SOME AISI–SAE STANDARD STEEL DESIGNATIONS

No.	*Name*	*Example*	*Nominal composition (wt%)*							
			C	*Mn*	*P (max)*	*S (max)*	*Si*	*Ni*	*Cr*	*Other*
10XX*	Plain carbon	1 015	0·15	0·45	0·04	0·05				
11XX	Free machining	1 119	0·19	1·15	0·04	0·3				
13XX	Mn	1 335	0·35	1·75	0·035	0·04	0·25			
3XXX	Ni Cr	3 140	0·4	0·8	0·04	0·04	0·3	1·0	0·6	
40XX	Mo	4 027	0·27	0·8	0·035	0·04	0·3			0·25 Mo
41XX	Cr Mo	4 118	0·18	0·8	0·035	0·04	0·3		0·5	0·1 Mo
43XX	Ni Cr Mo	4 320	0·2	0·55	0·035	0·04	0·3	1·8	0·5	0·25 Mo
46XX	Ni Mo	4 620	0·2	0·6	0·04	0·04	0·3	1·8		0·25 Mo
50XX	Cr	5 015	0·15	0·4	0·035	0·04	0·3		0·4	
61XX	Cr V	6 118	0·18	0·6	0·035	0·04	0·3		0·6	0·12 V
86XX	Ni Cr Mo	8 640	0·4	0·85	0·035	0·04	0·3	0·55	0·5	0·2 Mo
87XX	Ni Cr Mo	8 740	0·4	0·85	0·035	0·04	0·3	0·55	0·5	0·25 Mo
92XX	Si	9 260	0·6	0·85	0·035	0·04	2·0			
94BXX	Ni Cr Mo B	94B17	0·17	0·85	0·035	0·04	0·3	0·45	0·4	0·1 Mo 0·0005 B min

*The first two digits indicate the major alloying elements, the last two digits represent the carbon content.

TABLE 1.4

EFFECTS OF MAJOR ALLOYING ELEMENTS IN STEEL

Element	*Amount*	*Effect*
Aluminium	Small	Deoxidises, restricts grain growth, aids surface hardness in nitriding.
Boron	0·001–0·003%	Increases hardenability.
Carbon	less than 1·5%	Increases strength and hardness, decreases ductility and toughness.
Chromium	0·5–2·0%	Increases hardenability and strength.
	4·0–18·0%	Increases corrosion and oxidation resistance.
Cobalt	up to 12%	Improves cutting tool life at high temperatures, increases hardness.
Copper	0·1–0·4%	Improves corrosion resistance and strength.
Manganese	0·25–0·40%	Combines with sulphur to prevent brittleness.
	more than 1%	Improves hardenability, strength and hardness.
Molybdenum	0·2–5%	Improves hardenability, strength and hardness, enhances creep strength.
Nickel	2·0%–5·0%	Increases toughness especially at low temperatures.
	12·0–20·0%	Improves corrosion and oxidation resistance.
Silicon	0·2–0·7%	Deoxidises, increases strength.
	2·0%	Improves elastic properties for springs.
	more than 2·0%	Decreases losses in magnetising steel with alternating current.
Sulphur	0·08–0·15	Improves machinability.
Tungsten	up to 20·0%	Increases hardness at room and high temperatures, improves hardenability.
Vanadium	up to 5·0%	Increases strength while retaining ductility both at room and high temperatures.

Table 1.5

PROPERTIES OF SELECTED CARBON AND ALLOY STEELS IN THE NORMALISED CONDITION

AISI No.	*Yield strength (MN/m^2)*	*Tensile strength (MN/m^2)*	*Elongation (%)*	*Hardness (BHN)*	*Impact strength (Izod) (joule)*	*Machinability index*
1 015	329	430	37·0	121	115	50
1 020	352	448	35·8	131	118	65
1 030	350	529	32·0	149	93·6	65
1 040	380	599	28·0	170	65	60
1 050	434	760	20·0	217	27	50
1 060	427	788	18·0	229	13·2	
1 080	532	1 026	11·0	293	6·8	
1 095	508	1 029	9·5	293	5·4	
1 118	324	485	33·5	143	103	80
1 137	402·5	679	22·5	197	63·7	70
1 340	567	849	22·0	248	92	65
3 140	609	905	19·7	262	54·5	55
4 130	443	679	25·5	197	86	65
4 140	665	1 036	17·7	302	22·5	
4 320	471	805	20·8	235	73	55
4 340	875	1 299	12·2	363	16	45
4 620	372	583	29·0	174	133	60
4 820	492	767	24·0	229	110	50
5 140	480	805	22·7	229	38	60
5 160	539	971	17·5	269	10·8	55
6 150	625	954	21·8	269	35·5	45
8 630	436	660	23·5	186	95	
8 650	698	1 040	14·0	302	13·6	
8 740	616	943	16·0	269	17·6	
9 225	588	947	19·7	269	13·6	
9 310	579	921	18·8	269	119	

form or cut other materials. To perform its requirements efficiently, a tool steel is required to have excellent wear and abrasion resistance, high hardness at elevated temperatures and high toughness. The compositions, properties and uses of these steels will be discussed in detail in Chapter 14.

Cast irons are a group of materials that are basically ternary alloys of iron, carbon and silicon in which more carbon is present than can be retained in solid solution in austenite at the eutectic temperature (1403 K). The group of cast irons includes grey, white, malleable and nodular irons.

From Table 1.4 it is apparent that two or more alloying elements may produce similar effects. Thus it is possible to obtain steels with almost identical properties but with different chemical compositions. Some alloying elements are much more expensive than others, and some may be in short supply in certain countries, which can affect the cost of the different alloy steels. This fact should be kept in mind when selecting an alloy steel for a given application. In most cases the best steel to use is the cheapest one that can be satisfactorily heat treated to have the desired properties. An example of what can be achieved is the series of EX steels that have been designed by SAE to reduce the need for expensive alloying elements, notably Ni and Cr. Table 1.6 lists the composition of some EX steels and a comparison with the equivalent AISI–SAE steels of Table 1.3 readily shows the savings.

The group of low alloy steels that have yield strengths in excess of about 300 MN/m^2 are called high strength low alloy steels (HSLA). These steels usually contain from 0·05 to 0·33% C, and 0·2 to 1·65% Mn with small addition of Cr, Mo or Ni. HSLA steels are stronger than carbon steels and have higher toughness at low temperatures. Also, most HSLA grades are more resistant to atmospheric corrosion than are plain carbon steels.

Another group with even higher strength than HSLA are the ultra-high strength steels with yield strengths of about 1400 MN/m^2 or higher. Ultra-high strength steels differ widely in composition and in the way that their optimum strength is obtained, and they include some low, medium and high alloy steels, maraging steels and some stainless steels. Table 1.7 lists the composition and characteristics of some ultra-high strength steels.

Stainless steels constitute a large and widely used family of iron chromium alloys known for their corrosion resistance. This ability to resist corrosion is attributed to the self-healing and non-porous chromium oxide film that forms in the presence of oxygen. A minimum of 12% Cr is required for the formation of the film. Stainless steels are divided into three categories according to their microstructure: austenitic, ferritic and martensitic. Besides their non-rusting qualities, stainless steels exhibit a wide range of mechanical properties, e.g. austenitic grades show excellent performance at temperatures up to about 1000 K and retain most of their ductility and impact strength down to at least 80 K. Table 1.8 gives the compositions and strengths of representative stainless steels of the different categories. Iron base superalloys are capable of withstanding higher service temperatures than those permissible for stainless steels. These alloys can serve at temperatures up to about 1100 K and are mostly used in gas turbine parts like ducts, bolts, exhaust covers and tail cones.

Tool steels constitute a class of high-carbon alloy steels that are used to

TABLE 1.6

SOME EX STEELS AND EQUIVALENT AISI–SAE GRADES

EX No.	*Composition* (%)					*Equivalent AISI-SAE grade*
	C	*Mn*	*Cr*	*Mo*	*Other*	
12	0·38–0·43	0·75–1·00	0·25–0·40	0·05–0·10	0·20–0·40 Ni, 0·0005 B min	8 640
19	0·18–0·23	0·90–1·20	0·40–0·60	0·08–0·15	0·0005 B min	94B17
29	0·18–0·23	0·75–1·00	0·45–0·65	0·30–0·40	0·40–0·70 Ni	4 320
33	0·17–0·24	0·85–1·25	0·20 min	0·05 min	0·20 Ni min	4 027
36	0·38–0·43	0·90–1·20	0·45–0·65	0·13–0·20		8 640
52	0·38–0·43	0·95–1·25	0·25–0·40	0·05–0·10	0·20–0·40 Ni	8 640
54*	0·19–0·25	0·75–1·05	0·40–0·70	0·05 min		4 118

*All steels contain 0·035 P max, 0·04 S max, and 0·2–0·35 Si, except EX54 (0·35 Si max)

TABLE 1.7

COMPOSITION AND CHARACTERISTICS OF SOME ULTRA-HIGH STRENGTH STEELS

Steel	Composition (%)							Characteristics	Uses
	C	*Mn*	*Si*	*Cr*	*Ni*	*Mo*	*Other*		
Low alloy									
AISI 4130	0·3	0·5	0·3	0·95	—	0·2	—	Yield strength about 1800 MN/m^2.	Structure components for aircraft, axes, gears and shafts.
AISI 4340	0·4	0·7	0·3	0·8	1·8	0·25	—	Moderate cost, high hardenability.	
Medium alloy									
5 Cr Mo V	0·4	0·3	1·0	5·0	—	1·3	0·5 V	Air hardening. Yield strength about 1700 MN/m^2. Elongation about 5%.	Aircraft landing gear and structural components.
HY 130/150	0·1	0·75	0·25	0·5	5·0	0·5	0·07 V, 0·02 Al	Yield strength about 1050 MN/m^2. High toughness and good weldability.	Hulls for submarines, pressure vessels and thick walled cylinders.
High alloy									
HP9–4–30	0·3	0·2	0·1 max	1·0	7·5	1·0	4·5 Co, 0·1 V	Yield strength about 1540 MN/m^2. High toughness and good weldability.	Same as HY 130/150.

(*Contd.*)

TABLE 1.7 (*Contd.*)

Steel	*Composition* (%)							*Characteristics*	*Uses*
	C	*Mn*	*Si*	*Cr*	*Ni*	*Mo*	*Other*		
Maraging									
18 Ni (200)	0·03 max	0·1 max	0·1 max	—	18·0	3·25	8·5 Co, 0·2 Ti, 0·1 Al	Yield strength about 1400 MN/m².	ɔt working dies.
18 Ni (300)	0·03 max	0·1 max	0·1 max		18·5	4·9	9·0 Co, 0·65 Ti, 0·1 Al	Yield stength about 2100 MN/m².	Aircraft landing gear.
Stainless steel									
AISI 420	0·15 max	1·0 max	1·0 max	13·0	—	—	—	Combine high strength and corrosion resistance. Yield strength ranges from 1750 to 2000 MN/m².	Ducts for aircraft, nuclear reactor components equipment for food and chemical industries.
17–4 PH	0·07 max	1·0 max	1·0 max	16·5	4·0	—	4·0 Cu, 0·3 Nb		
17–4 PH	0·09 max	1·0 max	1·0 max	17·0	7·0	—	1·0 Al		
AISI 301	0·1	2·0 max	1·0 max	17·0	7·0	—	—		

TABLE 1.8

COMPOSITION AND PROPERTIES OF SELECTED STAINLESS AND HEAT-RESISTING STEELS

AISI	*Nominal composition* (%)						*Tensile strength*	*Yield strength*	*Elongation* (%)	*Hardness* (*BHN*)
	C	*Cr*	*Ni*	*Mn*	*Si*	*Other*	(MN/m^2)	(MN/m^2)		
Austenitic										
201	0·15	17	4·5	6·50	1·0	0·06 P, 0·03 S	805	385	55	185
301	0·15	17	7	2·0	1·0	0·045 P, 0·03 S	770	280	60	162
302	0·15	18	9	2·0	1·0	0·045 P, 0·03 S	630	280	50	162
304	0·08	19	9·5	2·0	1·0	0·045 P, 0·03 S	588	294	55	150
316	0·08	17	12	2·0	1·0	0·045 P, 0·03 S, 2·5 Mo	588	294	50	145
330	0·08	18·5	35·5	2·0	1·0	0·04 P, 0·03 S	630	266	45	150
Ferritic										
405	0·08	13	—	1·0	1·0	0·04 P, 0·03 S, 0·2 Al	490	280	30	150
430	0·12	17	—	1·0	1·0	0·04 P, 0·03 S	525	315	30	155
442	0·2	21·5	—	1·0	1·0	0·04 P, 0·03 S	560	315	20	185
Martensitic										
403	0·15	12·5	—	1·0	0·5	0·04 P, 0·03 S	525	280	35	153
416	0·15	13	—	1·25	1·0	0·06 P, 0·15 S, 0·6 Mo	525	280	30	153
431	0·2	16	2·0	1·0	1·0	0·04 P. 0·03 S	875	665	20	260
502	0·1	5	—	1·0	1·0	0·04 P, 0·03 S, 0·55 Mo	455	175	30	150
Iron–nickel base superalloys										
Incoloy 800	0·05	21	32·5	0·8	0·5	0·4 Ti, 0·4 Al, 46 Fe				

Cast irons are not identified by chemical analysis alone; the form of the excess carbon plays an important role in determining their type and behaviour. Table 1.9 compares the compositions and properties of the different types of cast iron. In grey iron the silicon content is high enough to cause the excess carbon to take the form of graphite flakes during solidification. The excess carbon is the basis for many of the desirable properties of grey iron, such as high fluidity, high damping capacity, low notch sensitivity and good machinability. Grey iron is the least expensive of all cast metallic materials and should always be considered first when a cast alloy is being selected. Another material should be chosen only when the mechanical or physical properties of grey iron are inadequate.

In white cast iron almost all the excess carbon is present as iron carbide, which makes it very hard but brittle. Annealing white iron transforms some or all of the iron carbide into graphite nodules and the resulting material is malleable cast iron. Among the useful characteristics of malleable iron are high toughness and excellent machinability and damping capacity.

Nodular cast iron, also known as ductile iron and spherulitic iron, is cast iron in which the excess carbon is present as tiny balls or spherulities, instead of flakes as in grey iron, or compacted aggregates as in malleable iron. The composition of unalloyed nodular irons is similar to that of grey irons but the spheroidal graphite is produced by the addition of one or more of the elements Mg, Ce, Ca, Li, Na and Ba. Mechanically, nodular cast iron is like mild steel but with the superior castability and machinability of cast irons.

1.5 NON-FERROUS ALLOYS

Non-ferrous metals and alloys are relatively more expensive than steels and are usually considered in the category of special materials to meet special needs. They are selected only when the relatively high cost per unit weight can be justified by one or more of their special properties, such as corrosion resistance, high electrical conductivity, high strength/weight ratio, or good bearing properties.

Aluminium alloys are the most important non-ferrous alloys and their selection is usually based on one or more of the following design considerations: 1, high strength/weight ratio; 2, resistance to atmospheric corrosion; 3, high electrical and heat conductivity, especially when light weight is important. The wrought aluminium alloys are identified by means of a four digit system developed by the Aluminum Association. The first

TABLE 1.9

COMPOSITION AND PROPERTIES OF SELECTED CAST IRONS

Specification No.	*Class*	*Composition* ($CE = \% C + 0{\cdot}3 (\% Si + P)$)	*Tensile strength* (MN/m^2)	*Average BHN*	*Yield strength* (MN/m^2)	*Elongation* (%)
Grey iron						
ASTM A48–74	20	$CE = 4{\cdot}34$	140	170	140	
	25	$CE = 4{\cdot}08$	175	190	175	
	35	$CE = 3{\cdot}77$	245	220	245	
	40	$CE = 3{\cdot}65$	280	225	280	
	50	$CE = 3{\cdot}45$	350	250	350	
	60	$CE = 3{\cdot}37$	420	270	420	
Nodular (ductile) iron						
ASTM A536	60–40–18	Chemical composition is	420	170	280	18
	65–45–12	subordinate to mechanical	455		315	12
	80–55–06	properties. However, the	560	215	385	6
	100–70–03	content of any chemical	700		490	3
	120–90–02	element may be specified by mutual agreement.	840	270	630	2
ASTM A395	60–40–18	$CE = 3{\cdot}77$	420	165	280	18
ASTM A476	80–60–03	$CE = 3{\cdot}8$–$4{\cdot}5$	560	201 min	420	3
Malleable iron						
ASTM	32 510	TC, 2·5%; Si, 1·3%; S, 0·11%;P, 0·18% max	350	110–145	230	10
A47 (ferritic)	35 018	TC, 2·3%; Si, 1·2%; S, 0·11%; P, 0·18% max	370		245	18
ASTM A220 (pearlitic)	40 010	TC, 2·3%; Si, 1·3%; S, 0·11%; P, 0·18% max	420	180–240	280	10

digit indicates the alloy type, the second digit indicates the alloy modifications, and the last two digits indicate the aluminium purity. Table 1.10 illustrates the designation system. Further identification of the alloy by temper was given in Table 1.1. Alloys in the 3 and 5 series are non-heat treatable and depend on solid solution and strain hardening for their strengthening. Alloys 2, 6 and 7 are heat treatable and although some of the alloys in the 4 series are heat treatable, most of them are used only for brazing sheets and welding wire. Table 1.11 gives some properties of representative aluminium alloys.

Cast aluminium alloys are also available in heat treatable and non-heat treatable grades and are designated by a three digit system as shown in Table 1.10. The relevant temper designations from Table 1.1 are also used for the heat treatable grades, e.g. T2, T4, T5 or T6. The mechanical strength and ductility of cast aluminium alloys are generally lower than those of the wrought alloys, as illustrated in Table 1.11.

Magnesium, with a specific gravity of only 1·74 is the lightest metal available for use by the engineer. Aluminium is the most commonly used alloying element and is used in amounts of up to about 10%. Zinc, rare earths, thorium, zirconium and manganese are also used as alloying elements as shown in Table 1.12. Although magnesium alloys are more expensive and weaker than aluminium alloys, their low density gives

TABLE 1.10

DESIGNATION SYSTEM FOR ALUMINIUM ALLOYS

I Wrought alloys	
1XXX	Aluminium 99% and greater, e.g. 1060 (99·6% Al)
2XXX	Two phase Cu alloys, e.g. 2014 (4·4% Cu, 0·8% Si, 0·8% Mn)
3XXX	One phase Mn alloys, e.g. 3003 (1·2% Mn)
4XXX	Two phase Si alloys, e.g. 4032(12·5% Si, 1% Mg, 0·9% Cu, 0·9% Ni)
5XXX	One phase Mg alloys, e.g. 5052, (2·5% Mg, 0·25% Cr)
6XXX	Two phase Mg–Si alloys, e.g. 6061 (0·6% Si, 1% Mg, 0·25% Cu, 0·25% Cr)
7XXX	Two phase Zn alloys, e.g. 7075 (5·6% Zn, 2·5% Mg, 1·5% Cu, 0·3% Cr)
II Cast alloys	
2XX	Al–Cu alloys, e.g. 208·0 (4% Cu, 3% Si, 1·2% Fe, 1·0% Zn)
3XX	Al–Si–Cu alloys, e.g. 333·0 (9% Si, 3·5% Cu, 1·0% Fe, 1·0% Zn)
4XX	Al–Si alloys, e.g. B443·0 (5% Si)
5XX	Al–Mg alloys, e.g. 520·0 (10% Mg)
7XX	Al–Zn alloys, e.g. A712·0 (6·5% Zn, 0·7% Mg)
8XX	Al–Sn alloys, e.g. 850·0 (6·5% Sn, 1% Cu, 1% Ni)

TABLE 1.11

COMPOSITION AND PROPERTIES OF SELECTED ALUMINIUM ALLOYS

Alloy	*Temper*	*Nominal composition* (%)	*Tensile strength* (MN/m^2)	*Yield strength* (MN/m^2)	*Elongation* (%)	*Hardness* (*BHN*)
Wrought alloys						
1 060	0	99·6 + Al	70	28	43	19
	H18		133	126	6	35
2 014	0	4·4 Cu, 0·8 Si, 0·8 Mn, 0·4 Mg	189	98	18	45
	T6		490	420	13	135
3 003	0	1·2 Mn	112	42	40	28
	H18		203	189	10	55
4 032	T6	12·5 Si, 1·0 Mg, 0·9 Cu, 0·9 Ni	385	322	9	120
5 052	0	2·5 Mg, 0·25 Cr	196	91	30	47
	H38		294	259	8	77
6 061	0	1·0 Mg, 0·6 Si, 0·25 Cu, 0·25 Cr	126	56	30	30
	T6		315	280	17	95
7075	0	5·5 Zn, 2·5 Mg, 1·5 Cu, 0·3 Cr	231	105	16	60
	T6		581	511	11	150
Casting alloys						
208·0	Sand-Cast	4 Cu, 3 Si	147	98	2·5	55
356·0	T51	7 Si, 0·3 Mg	175	140	2	60
	T6		231	168	3·5	70
B443·0	Sand-Cast	5 Si	133	56	8	40
	Die-Cast		231	112	9	50
520·0	T4	10 Mg	336	182	16	75
850·0	T5	6·5 Sn, 1 Cu, 1 Ni	161	77	10	45

TABLE 1·12

COMPOSITION AND PROPERTIES OF SELECTED MAGNESIUM ALLOYS

Alloy	*Temper*	*Nominal composition* (%)	*Tensile strength* (MN/m^2)	*Yield strength* (MN/m^2)	*Elongation* (%)	*Hardness* (*BHN*)
Wrought alloys						
AZ31B	0	3·0 Al, 0·2 Mn, 1·0 Zn	224	115	11	56
	H24		255	165	7	73
AZ61A	F	6·5 Al, 0·15 Mn, 1·0 Zn	266	140	8	55
HK31A	H24	0·7 Zr, 3·2 Th	235	175	4	57
HM21A	T8	0·8 Mn, 2·0 Th	224	140	6	55
ZK40A	T5	4·0 Zn, 0·45 Zr	280	250	4	60
Casting alloys						
AM100A	F	10·0 Al, 0·1 Mn	140	70	6	53
	T4		238	70	6	52
	T6		238	105	2	52
AZ63A	F	6·0 Al, 0·15 Mn, 3·0 Zn	182	77	4	50
	T6		238	112	3	73
EZ33A	T5	2·6 Zn, 0·7 Zr, 3·2 Re	140	98	2	50
HK31A	T6	0·7 Zr, 3·2 Th	189	91	4	55
HZ32A	T5	2·1 Zn, 0·7 Zr, 3·2 Th	189	91	4	55
ZK61A	T6	6·0 Zn, 0·8 Zr	280	182	5	70

cost/volume and strength/weight ratios that are competitive with aluminium alloys. Magnesium alloys are thus used in a wide variety of applications in the aerospace and aircraft fields. Other industrial uses include storage tanks, portable tools and moving parts in printing and textile equipment. In the ASTM designation system for magnesium alloys, the first two letters indicate the major alloying elements, followed by numbers representing the percentage of the two elements present, as shown in Table 1.12. The Mg–Al–Zn alloys (AZ) are the largest group of magnesium alloys and contain up to 10% aluminium and not more than 1·5% zinc. These alloys can be precipitation hardened and are generally the strongest of the magnesium alloys. Magnesium alloys are very easy to machine. The casting alloys have excellent castability, and in die-castings, wall thicknesses of about 1 mm are possible.

Titanium metal is classified as a light metal with a specific gravity of 4·5. Like iron, titanium is allotropic and exists as CPH, α-phase, at temperatures up to about 1158 K and then transforms to BCC, β-phase, which is stable up to the melting point (1941 K). This allotropic transformation is affected by the addition of alloying elements. For example, aluminium and carbon stabilise the α-phase, i.e. raise the α to β transformation temperature, while copper, chromium, iron, molybdenum and vanadium stabilise the β-phase. The mechanical properties of titanium alloys are closely related to these allotropic phases, e.g. β-phase is much stronger but more brittle than α-phase. Therefore these alloys can be classified as α, β and α + β. Besides their relatively light weight, titanium alloys are also strong, which gives them the highest strength/weight ratio among structural alloys. This exceptional property is maintained from 22 K up to 800 K. Titanium alloys also have excellent corrosion resistance to atmospheric and sea environments as well as to a wide range of chemicals.

α alloys contain alloying elements like Al, Sn, Nb and Zr in amounts of up to about 10%. They are non-heat treatable with a good combination of strength, toughness and weldability. They are also stable in the range 22 K up to 800 K. β alloys are very strong but their lack of toughness limits their use industrially. α + β alloys combine the characteristics of the α and β alloys so that they are stronger than the α alloys but less tough and more difficult to weld. The α + β alloys are heat treatable by ageing treatments. Table 1.13 lists a representative sample of titanium alloy compositions, properties and applications.

Copper and copper alloys are generally known for high thermal and electrical conductivity, high corrosion resistance and high ductility and strength. The fields of greatest use for pure copper are as electrical

TABLE 1.13

PROPERTIES AND APPLICATIONS OF SELECTED WROUGHT TITANIUM ALLOYS

Alloy	*Tensile strength* (*MN/m*^{2})	*Yield strength* (*MN/m*^{2})	*Elongation* (%)	*Hardness* (*RC*)	*Application*
Alpha alloys					
5 Al, 25 Sn	875	819	16	36	Weldable, aircraft engine compressor
8 Al, 1 Mo, 1 V	1 029	945	16	—	blades and ducts, steam turbine blades.
Alpha + beta alloys					
3 Al, 2·5 V	700	595	20	—	Aircraft hydraulic tubes.
6 Al, 4 V	1 008	939	14	36	Rocket motor cases, blades and discs for turbines.
7 Al, 4 Mo	1 120	1 050	16	38	Airframes and jet engine parts.
6 Al, 2 Sn, 4 Zr, 6 Mo	1 288	1 190	10	42	Components for advanced jet engines.
10 V, 2 Fe, 3 Al	1 295	1 218	10	—	Airframe structures requiring toughness and strength.
Beta alloys					
13 V, 11 Cr, 3 Al	1 239	1 190	8	—	High strength fasteners.
8 Mo, 8 V, 2 Fe, 3 Al	1 330	1 260	8	40	Aerospace components.
3 Al, 8 V, 6 Cr, 4Mo, 4 Zr	1 470	1 400	7	42	High strength fasteners.
11·5 Mo, 6 Zr, 4·5 Sn	1 407	1 337	11	—	High strength sheets for aircraft.

conductors when in the form of wire, and as heat exchangers when in the form of tubes. The most widely used copper alloys are brasses (Cu + Zn), tin bronzes (Cu + Sn), phosphor bronzes (Cu + Sn + P), aluminium bronzes (Cu + Al), manganese bronzes (Cu + Zn + Sn + Mn) and copper nickel (Cu + Ni). The alloys can be broadly divided into wrought alloys and cast alloys. In the wrought class, single phase alloys are usually strengthened by solid solution or strain hardening, and multiphase alloys are strengthened by precipitates or second-phase dispersions. Table 1.14 lists the composition and properties of selected wrought copper alloys. The cast alloys offer a wider range of structures and are more tolerant of impurities than wrought alloys, e.g. alloys containing a high percentage of lead are not suited for hot working, but can be easily cast. Typical compositions and properties of some cast copper alloys are given in Table 1.15.

Nickel is an important metal that is used as an alloying element in ferrous and non-ferrous alloys and as the base metal for a number of nickel alloys. In general, nickel alloys are stronger, tougher and harder than most other non-ferrous alloys and many steels, but they are more costly than aluminium alloys and most ferrous alloys. A commonly used group of alloys is the nickel–copper group known commercially as the Monel alloys. Monel metal has an average composition of 67% Ni, 28% Cu and 5% Fe. This alloy resists corrosion from most chemicals and is a competitor to stainless steel in the chemical, marine, power equipment, food service, petroleum and paper industries. Some nickel–chromium alloys are used in cryogenic applications as they have outstanding strength, toughness and ductility at low temperatures, as will be discussed in Chapter 19. However, the most important use for nickel-base alloys is at high temperatures in the range 1350 K to 1500 K. The high temperature alloys (nickel-base superalloys) are basically Ni–Cr alloys with additions of Co, Mo, Ti or Al and depend mostly on precipitation hardening for strengthening. The composition and properties of representative superalloys will be discussed in Chapter 18.

Cobalt is used as an alloying element in many important steels and as a base metal for a number of important superalloys. These superalloys contain 20–30% chromium for oxidation resistance and many of them contain tungsten, carbon, nickel, aluminium and minor additions of manganese, silicon, titanium and molybdenum. Most of these alloying elements form hard carbides in a fairly hard matrix and give excellent abrasion and wear resistance. The composition, properties and characteristics of representative cobalt-base superalloys will be discussed in

TABLE 1.14

COMPOSITION AND PROPERTIES OF SELECTED WROUGHT COPPER ALLOYS

Alloy	*Nominal composition*	*Treatment*	*Tensile strength* (MN/m^2)	*Yield strength* (MN/m^2)	*Elongation* (%)	*Rockwell hardness*
Gilding, 95%	95 Cu, 5 Zn	Annealed	245	77	45	RF 52
		hard	392	350	5	RB 64
Red brass, 85%	85 Cu, 15 Zn	Annealed	280	91	47	RF 64
		hard	434	406	5	RB 73
Cartridge brass, 70%	70 Cu, 30 Zn	Annealed	357	133	55	RF 72
		hard	532	441	8	RB 82
Muntz metal	60 Cu, 40 Zn	Annealed	378	119	45	RF 80
		half-hard	490	350	15	RB 75
High-leaded brass	65 Cu, 33 Zn, 2 Pb	Annealed	350	119	52	RF 68
		hard	518	420	7	RB 80
Phosphor bronze, 5%	95 Cu, 5 Sn	Annealed	350	175	55	RB 40
		hard	588	581	9	RB 90
Phosphor bronze, 10%	90 Cu, 10 Sn	Annealed	483	250	63	RB 62
		hard	707	658	16	RB 96
Cupro-nickel, 30%	70 Cu, 30 Ni	Annealed	385	126	36	RB 40
		cold rolled	588	553	3	RB 86
Nickel silver	65 Cu, 23 Zn,	Annealed	427	196	35	RB 55
(German silver)	12 Ni	hard	595	525	4	RB 89
High silicon bronze	96 Cu, 3 Si	Annealed	441	210	55	RB 66
		hard	658	406	8	RB 93
Aluminium bronze	95 Cu, 5 Al	Annealed	420	175	66	RB 49
		cold rolled	700	441	8	RB 94
Aluminium bronze (2)	81·5 Cu, 9·5 Al,	Soft	630	—	12	—
	5 Ni, 2·5 Fe, 1 Mn	hard	735	420	12	RB 105
Beryllium copper	97·9 Cu, 1·9 Be,	Annealed	490	—	35	RB 60
	0·2 Ni or Co	HT (hardened)	1400	1050	2	RC 42

TABLE 1.15

COMPOSITION AND PROPERTIES OF SELECTED CAST COPPER ALLOYS

Alloy	*Nominal composition* (%)	*Tensile strength* (MN/m^2)	*Yield strength* (MN/m^2)	*Elongation*	*Hardness* (*BHN*)
Tin bronze (alloy 1A)	88 Cu, 10 Sn, 2 Zn	350	150	35	80
Tin bronze, SAE 65	89 Cu, 11 Sn	400	200	35	95
Leaded tin bronze, (alloy 2B)	87 Cu, 8 Sn, 1 Pb, 4 Zn	280	150	25	68
High-lead tin bronze (alloy 3D)	78 Cu, 7 Sn, 15 Pb	200	112	15	55
Manganese bronze (alloy 8C)	64 Cu, 26 Zn, 3 Fe, 5 Al, 4 Mn	805	490	15	210
Aluminium bronze (alloy 9C)	85 Cu, 4 Fe, 11 Al	630	231	12	160
Aluminium bronze (Propeller bronze)	82 Cu, 4 Fe, 9 Al, 4 Ni, 1 Mn	630	266	20	150
Nickel silver (alloy 11B)	66·5 Cu, 5 Sn, 1·5 Pb, 2 Zn, 25 Ni	420	210	20	145
Silicon bronze (alloy 12A)	87 Cu, 4 Si, 1 Sn, 4 Zn, 2 Fe, 1 Al, 1 Mn	400	119	15	120
Silicon brass (alloy 13B)	81 Cu, 4 Si, 15 Zn	490	224	24	134

Chapter 18. Cobalt is also a significant element in low expansion alloys, e.g. 53–54·5% Co, 36·5–37% Fe and 9–10% Cr alloy, which has an exceedingly low coefficient of expansion, at times negative, over the range from 273 K to 373 K. Cobalt–chromium-base alloys are used in dental and surgical applications because they are not attacked by body fluids, as will be discussed in Chapter 17.

Refractory metals are a group of metals with melting points above 1900 K. The most important metals in this group are tungsten, molybdenum, tantalum and niobium. At temperatures above 1350 K these metals and their alloys generally have higher strengths than other high temperature alloys. However, refractory metals are generally susceptible to oxidation at elevated temperatures. Consequently they must be protected with high temperature coatings. Also these refractory metals and their alloys involve difficult problems in fabrication, especially when the product is in the sheet form. The most advanced applications of refractory metals and their alloys are in the manufacture of rocket motors and turbojet engines. Table 1.16 lists some relevant properties of representative refractory metals and alloys.

Zinc is a relatively inexpensive metal with a relatively low melting point of 692·5 K. Over 40% of the zinc produced is used to assist the control of corrosion of iron and steel by galvanising. Another important use for zinc is as an alloying element in brasses. Zinc is also used in dry batteries, photo-engraving and as the base metal for some important die-casting alloys. The two alloys presently in use are known as ASTM AG40A (XXIII) and AG41A (XXV) or, respectively, SAE 903 and 925. The composition and properties of these alloys are given in Table 1.17. Alloy AG40A has better ductility and retains its impact strength better at elevated temperatures. However, alloy AG41A is harder and stronger and has better castability. Both alloys can be strengthened by ageing. At room temperature the impact strength of zinc alloy die-castings is much higher than that of either aluminium or magnesium die-castings or iron sand-castings. Some wrought zinc alloys exhibit superplastic behaviour, e.g. Zn containing 22% Al.

Tin, lead, bismuth, antimony, cadmium and indium and their alloys can be grouped together as low-melting materials. The common characteristics of these materials, besides their low melting points, are low strength and hardness, high ductility and relatively high corrosion resistance. These characteristics give most of these materials wide use as bearing materials and soldering alloys. Lead-base and tin-base bearing alloys, babbitts, will be discussed in Chapter 15. A major use for pure tin is for coating other metals. It is applied to steel and copper to provide corrosion resistance, non-

TABLE 1.16

COMPOSITION AND PROPERTIES OF SELECTED REFRACTORY METALS AND ALLOYS

Alloy	*Nominal additions (%)*	*Test temperature (K)*	*Tensile strength at test temperature (MN/m^2)*	*10 h rupture stress (MN/m^2)*
Niobium alloys				
Unalloyed Nb	None	1 366	70	38
SCB291	10 Ta, 10 W	1 366	224	63
C129Y	10 W, 10 Hf, 0·1 Y	1 590	182	105
FS85	28 Ta, 11 W, 0·8 Zr	1 590	161	84
Molybdenum alloys				
Unalloyed Mo	None	1 366	182	102
TZM	0·5 Ti, 0·08 Zr, 0·015 C	1 590	371	154
WZM	25 W, 0·1 Zr, 0·03 C	1 590	504	105
Tantalum alloys				
Unalloyed Ta	None	1 590	60	17.5
Ta – 10 W	10 W	1 590	350	140
T-222	9·6 W, 2·4 Hf, 0·01 C	1 590	280	266
Tungsten alloys				
Unalloyed W	None	1 922	175	48
W – 2 ThO_2	2 ThO_2	1 922	210	126
W – 15 Mo	15 Mo	1 922	252	84
GE 218	Doped	1 922	406	—

toxicity and white colour. Tin is also an important alloying element in bronzes. In the metallic form, lead finds an important application in building construction as pipe, sheet and solder. Lead is also used in important alloys like type metal, antimonial lead for storage batteries, bearing alloys, foil and collapsible tubes. Lead is also used as shielding against X-rays and gamma radiation. The major use for bismuth, antimony, cadmium and indium is as alloying elements, especially to the low-melting alloys, as shown in Table 1.18.

The precious metals include gold, silver and the platinum group metals (platinum, palladium, rhodium, ruthenium, iridium and osmium). The precious metals, except for silver, are the most expensive known metals. In spite of their high cost, these metals find considerable use in industry because of their chemical inertness. Table 1.19 gives some representative alloys and some uses for precious metals in industry.

TABLE 1.17

COMPOSITION AND PROPERTIES OF ZINC DIE-CASTING ALLOYS

	Zamak 3 ASTM AG40A (XXIII)	*Zamak 5 ASTM AG41A (XXV)*
Composition (%)		
Copper	0·25	0·75–1·25
Aluminium	3·5–4·3	3·5–4·3
Magnesium	0·02–0·05	0·03–0·08
Iron, max	0·100	0·100
Lead, max	0·005	0·005
Cadmium, max	0·004	0·004
Tin, max	0·003	0·003
Zinc	Remainder	Remainder
Properties		
Tensile strength (MN/m^2)	287	329
Compressive yield strength (MN/m^2)	420	609
Elongation (%)	10·00	7·00
Charpy impact (joule)	58·3	66
Hardness (BHN)	82	91
Specific gravity	6·6	6·7

1.6 MANUFACTURING BY CASTING

Casting is a versatile manufacturing process, since it can be applied to a wide variety of alloys, shapes and product sizes. It is the only way in which bodies of complex internal shape can be economically made, and it is the most economical way of forming the more complex high temperature alloys. Although the mechanical performance of cast structures is usually inferior to wrought structures, extremely strong cast alloys have been produced by directional solidification, e.g. *in situ* composites. Several casting processes are industrially available and the differences between them can be expressed in terms of the possible product size, surface finish, tolerance on dimensions, minimum section thickness, production rate, mould or pattern cost and scrap loss. Tables 1.20 and 1.21 compare these parameters for some of the commercially available casting processes.

Sand-casting is the cheapest and most versatile process as it can be employed for the widest range of alloys and product shapes and sizes, but it is the least accurate, as shown in Table 1.20. Shell moulding, in which a

TABLE 1.18

COMPOSITION, MELTING TEMPERATURES AND APPLICATIONS OF LOW-MELTING ALLOYS

Composition (%)					*Melting point (K)*	*Applications*
Bi	*Pb*	*Sn*	*Cd*	*Other*		
Eutectic alloys						
44·7	22·6	8·3	5·3	19·1 In	319.8	Lens buffing and grinding, dental models.
55·5	44·5	—	—	—	397	Gem cutting, electroforming, founding.
58·00	—	42·00	—	—	411·5	Embossing dies, founding.
49·00	18·00	12·00	—	21 In	331	Locator for precision parts for testing.
Non-eutectic alloys						
42·5	37·7	11·3	8·5	—	431–467	Locator for needles in lace and textile machines.
33·33	33·34	33·33	—	—	467–562	Bushing for large bearings.
48·0	28·5	14·5	—	9 Sb	490–713	Bushing for small bearings, anchors, repairing broken dies.
40·0	—	60·00	—	—	554–611	Founding, electroforming, gem cutting.

resin coated sand is formed into a thin walled mould by contact with a heated pattern, fits in between sand-casting and precision methods. Die-casting gives higher dimensional accuracy and an improved surface finish characteristic of metallic dies. The accuracy and surface finish of the cast product depends on the pressure used to force the metal into the die mould. The high price of the metal mould in die-casting makes it economical only when numbers in excess of about 5000 are to be cast.

Investment casting, known as the lost wax process, gives the designer an almost unlimited freedom in choosing his shapes and materials in sizes varying from a few grams to several kilograms. The ability to hold close tolerances enables final machining processes to be kept to a minimum or eliminated. This latter property enables alloys that are difficult to machine to be made into precise shapes.

Besides the above process variables, the quality of cast products is also affected by the properties of the cast alloy. Fluidity is the ability of a molten alloy to fill a mould cavity completely, and castability is the ease with which

TABLE 1.19

COMPOSITION, PROPERTIES AND USES OF SELECTED PRECIOUS METALS AND ALLOYS

Alloy	*Tensile strength (MN/m^2)*	*Hardness (VHN)*	*Characteristics*	*Use in industry*
Pure silver	127	25	Corrosion resistance, high conductivity.	Electrical conductor at red heat. Silver plating, reflectors, catalyst.
Coin silver (90% Ag, 10% Cu)	350	100	Harder than silver.	Coins, electrical contacts.
Dental amalgam (33% Ag, 52% Hg, 12·5% Sn, 2% Cu, 0·5% Zn)	330	90	Liquidus temperature 348 K.	Silver dental fillings.
Gold (70%), platinum (30%)	650	150	High corrosion resistance.	Spinnerets in rayon production.
Platinum (99·85%)	155	40	High melting point, ductility, oxidation resistance, insolubility.	Resistance thermometers, thermocouples, precision electrical meters, spinnerets for synthetic fibres.
Platinum (3·5–40% Rh)	400–1300	120–300	Stronger than pure Pt.	Spinnerets for rayon, furnace windings, thermocouples.
Platinum (4–8% W)	570–2100	100–300	Excellent wear resistance, low electron emission.	Special spark plug electrodes, grids in power tubes for radar, potentiometer wire.
Palladium (99·85%)	150	38	Resists tarnishing.	Contacts in electrical relays.
Palladium (60%), copper (40%)	1350	—	Harder and cheaper than pure Pd.	Electrical contacts, slip rings.
Iridium	650	350	The most corrosion resistant element.	Crucibles for high temperature reactions.
Osmium	—	800	Hard, wear and corrosion resistant.	Fountain pen nibs, phonograph needles, instrument pivots.

TABLE 1.20

CHARACTERISTICS OF CASTING AND POWDER METALLURGY METHODS

Process	*Description*	*Materials cast*	*Advantages*	*Limitations*
Sand-casting	Mould is either green or dry sand and is broken to remove casting.	Ferrous alloys, superalloys, light- and low-melting alloys.	Extreme shape complexity possible, no limits on size, low tooling cost.	Difficult to get close tolerances or good surfaces, large radii preferred.
Shell moulding	Mould is resin-coated sand and is broken to remove casting.	Same as sand-casting.	High production rates, better tolerances and surface finishes than sand-casting.	More expensive tooling, minimum core usage preferable.
Investment casting	Mould is fired refractory slurry.	Same as sand-casting.	Close tolerances, good surface finish, no parting line or flash.	Limited sizes, expensive patterns high labour cost.
Low pressure die-casting	Metallic mould, metal fed under gravity.	Al, Mg, Zn, Pb, Sn, some copper alloys.	Good surface finish and high tolerances.	High die costs, limited to melting points less than 1200 K, limited to small simple parts.
Pressure die-casting	Molten metal forced into a metallic mould at high pressure.	Same as low pressure die-casting.	Better quality than low pressure die-casting.	Higher cost than low pressure die-casting.
Centrifugal casting	Rotating mould, centrifugal forces throw metal against mould wall.	Same as sand-casting.	Produces large cylindrical parts with thin walls.	Limited shapes, expensive equipment.
Powder metallurgy	Cold pressing and sintering of powders or hot pressing.	All metallic materials.	Density and porosity can be controlled, low losses, close tolerances.	Limited shapes and sizes, holes not in direction of pressing cannot be produced.

TABLE 1.21

COMPARISON OF SOME FEATURES OF CASTING AND POWDER METALLURGY METHODS

Process	*Commercial sizes (kg)*	*Commercial surface finish (μm)*	*Tolerance (mm/mm)*	*Porosity rating*	*Minimum section thickness (mm)*	*Relative production rate*	*Relative mould or pattern cost*
Sand-casting	0·02–10^5	5–15	± 0·03–0·2	fair	3–5	1	1
Shell moulding	0·02–10·0	3–20	± 0·004	good	3–5	4	7
Investment casting	0·01–5·0	0·4–3	± 0·003	very good	0·4–1·0	6	7
Low pressure die-casting	0·02–5·0	1–3	± 0·01–± 0·05	very good	1–3	4·5	10
Pressure die-casting	0·01–5·0	0·4–2.5	± 0·001–± 0·05	excellent	0·5–1·5	10	15
Centrifugal casting	up to 10^3	0·6–5	± 0·01–± 0·05	excellent	1·5	3	5
Powder metallurgy	0·01–5·0	0·4–2·5	± 0·002 in radial dimensions	variable	0·8	8	12

an alloy responds to ordinary foundry practice without undue attention to gating, risering, melting and mould condition. The shrinkage of the alloy on solidification and cooling to room temperature is also an important parameter in determining the quality of cast products. High shrinkage necessitates more careful design to promote directional solidification, reduce abrupt changes in cross section, liberalise fillets, and permit the best placement of gates and risers, thus avoiding internal shrinkage.

Metallic materials are cast either to produce a semi-finished component or merely to obtain a more convenient form for subsequent working. The as cast structures usually suffer from segregation or coring and in some cases porosity is also present. These defects make cast structures relatively weak and brittle. However, this is not usually serious in unstressed components or if the material is to be subsequently worked. If a stressed component is to be produced by casting, special effort should be made to obtain porosity free structures and some form of stress relief and homogenisation treatment is usually recommended. Table 1.22 gives typical mechanical properties of some common cast alloys. In the cases of mass production and automatic machining operations, the cast material should have highly reproducible properties with respect to hardness, machinability and freedom from defects. These requirements can be more easily achieved in some alloys than in others. Another factor which should be considered when designing a casting is the effect of the cooling rate on the structure and mechanical properties. Sensitive materials can become weaker with slower cooling rates, and a section that has been made excessively thick in order to bear more load can be a source of potential weakness.

Powder metallurgy techniques, although different from normal casting, yield similar products and are subject to the same design limitations. Powder metallurgy products can be made either by cold pressing followed by sintering and sizing, or by combining pressure and heat in hot pressing. Powder preforms can also be prepared for further forming processes. Powder metallurgy techniques are compared with the different casting processes in Tables 1.20 and 1.21. Die life is a critical factor in assessing the processes' suitability for high-speed automated production. The presence of porosity in sintered parts reduces their strength, ductility and impact strength relative to high quality forgings and castings. Hence methods which can minimise porosity will produce competitive components, e.g. hot isostatic compaction. As a rule, powder metallurgy is not considered competitive for quantities of less than 20 000 pieces. The capacities of ordinary production equipment limit the process to relatively simple parts with cross sectional area of about 25 cm^2 or less, and weight of less than about 10 kg, although some larger parts have been made.

TABLE 1.22

PROPERTY RANGE OF SOME COMMON CASTING ALLOYS

Casting alloy group	*Tensile strength range* (MN/m^2)	*Yield strength range* (MN/m^2)	*Elongation range* (%)	*Hardness range* (*BHN*)	*Endurance limit range* (MN/m^2)
Ferrous alloys					
Plain carbon steel	350–700	175–350	15–35	110–210	140–350
Low alloy steel	630–1400	420–1400	5–20	170–370	250–700
Grey iron	140–560	140–560	0–3	130–350	70–210
Nodular iron	420–1120	250–945	1–25	140–450	168–560
Malleable iron	350–840	140–290	1–20	110–290	140–420
Non-ferrous alloys					
Aluminium alloys	140–350	50–210	1–20	40–80	40–140
Magnesium alloys	140–300	70–200	2–10	50–80	40–120
Zinc alloys	280–340	100–200	5–10	80–100	80–140
Copper alloys	180–900	100–300	10–40	50–210	50–350

1.7 MANUFACTURING BY WORKING

The ability to deform plastically is a very important property of metallic materials as it enables them to be both strong and tough in service and also to be shaped in the solid state by various mechanical working processes. Metallic materials are generally worked for two main reasons. First, to produce shapes that would be difficult or expensive to produce by other means; these may range from sheets or wire, to more complex shapes such as I-beams and rails. The second reason is that the mechanical properties of metallic materials are usually improved by mechanical working. The improvement is achieved through grain refinement by recrystallisation, directional control of flow lines, homogenisation of cored structure, breaking-up and distribution of structural irregularities and inclusions, and closing up and welding of porosity. All metal working processes have a common feature of grain flow and this must be taken into account by the designer, since mechanical properties are best in the direction of flow, and local variations can be advantageously exploited.

Metal forming above the recrystallisation temperature but below the melting point is called hot working, while forming below the recrystallisation temperature is cold working. For most metallic materials, the minimum hot working temperature is about 0·5 *Tm*, where *Tm* is the absolute melting temperature. Generally, hot worked materials are soft and ductile while cold worked materials are strain-hardened and less ductile. Therefore hot working processes are usually performed in the semi-finishing stages of production and cold working processes in the finishing stages in order to take advantage of the improved strength, the closer tolerances and the improved surface finish that result. Table 1.23 lists the characteristics of common industrial working processes.

Forging can be divided into two groups depending on the use of an open or a closed die. Open die forging is used for large components which have to be made in small numbers and, since the tooling costs are low compared with a closed die technique, it is cheaper. Besides being less accurate, open die forging needs skilled labour and has a slow rate of production. Closed die forging gives greater design flexibility, enabling more complex forms with closer tolerances to be made; and with automation, labour costs are reduced and reproducibility is improved. It should be noted that specifying more accuracy and better surface finish for a forged component will necessitate using higher pressures on the tool, which reduces tool life and hence the viability of the process. Because of the high cost of forging dies, at least 30 000 components are usually needed before the process becomes economic.

TABLE 1.23

CHARACTERISTICS OF COMMON INDUSTRIAL WORKING PROCESSES

Process	*Description*	*Materials worked*	*Advantages*	*Limitations*
Open die forging	Material heated to hot working temperature and hammered to shape with little lateral confinement.	All metallic wrought alloys.	Inexpensive simple tools. Economic for small production.	Limited to simple shapes and large tolerances. Slow production rates. Requires high labour skill.
Closed die forging	Same as open die forging except that the forged material is confined all round by die cavity.	All metallic wrought alloys.	High production rates and good tolerance and reproducibility.	Higher tool costs. Limited product size.
Rolling	Material is passed through rotating rolls and takes the form of the roll gap.	Almost all metallic wrought alloys.	High rates of production. High tolerance in cold rolling. The only practical way to produce wide, thin sheets.	High capital cost. Not used for hollow shapes.

Extrusion	Material is pushed through a die orifice shaped in the required form.	Almost all metallic wrought alloys but more common for non-ferrous.	Even relatively brittle materials can be extruded. High rates of production.	High capital and tooling cost. Difficult to use for very strong materials.
Blanking, stamping, and deep drawing	Sheet metal is sheared or pressed into dies of required shape.	All sheet metallic alloys.	High production rates, good surfaces and tolerance. Large variety of shapes and sizes.	High tool costs. Limited to sheet metals. High scrap ratio.
Wire and tube drawing	Material is drawn through dies at room temperature to reduce outer diameter and sometimes wall thickness in tubes.	Wrought materials with enough ductility at room temperature.	The only way to produce fine wires. Good tolerance and surface finish.	Uniform cross sections only.
Explosive forming	Shock waves deform material to required shape.	All wrought alloys	No size limits, high production rates can improve formability.	Low production rates. Skilled labour required.

Rolling is used to produce a large proportion of flat products like plate, sheet and strip, as well as structural shapes like angles, rails and channels. A large quantity of rods and bars are also produced by rolling. The structure and properties of hot rolled products can be controlled by controlling the rolling schedule and finishing temperature. In cold rolled products, the directionality of properties, degree of strain-hardening, surface finish and dimensional tolerance are the main parameters which can be controlled by varying the production techniques.

Extrusion is usually used in the manufacture of non-ferrous metals and alloys, as complicated shapes of constant cross sectional area can be produced easily, either hot or cold. Impact extrusion is used to produce a large proportion of closed-end, thin-wall tubes from soft alloys. Forging and extrusion processes can be combined to produce a wide variety of shapes and products.

Wire and tube drawing usually follow hot rolling or extrusion in the production sequence, while sheet metal work usually follows cold rolling and consists of bending, shearing and deep drawing operations. These processes involve expensive tooling, and economic lot sizes should be kept in mind by the designer.

1.8 JOINING PROCESSES

Joining is the process of permanently, or sometimes temporarily, bonding or attaching materials to each other. Nearly all engineering structures are constructed from various parts that have to be joined together into units. Joining processes can be classified into welding, brazing and soldering, adhesive bonding and mechanical fastening.

Welding involves metallurgically joining the parts together to produce essentially a single piece. Welding processes may be classified under the headings of forge welding, fusion welding and pressure welding. Brazing and soldering are similar to welding and are included in Table 1.24 where the different industrial methods of welding are given. Welding processes involve most of the important metallurgical processes such as melting, alloying, casting, solidification, hot and cold working and heat treatment. In fusion welding processes, the molten filler metal solidifies quite rapidly by heat conduction into the metal adjacent to the weld. Columnar grains are usually present in the weld bead while the base metal closest to it undergoes a considerable overheating and grain growth. This latter area, the heat affected zone, is usually a source of failure in welded components. Thermal contrac-

TABLE 1.24

COMMON WELDING OPERATIONS

Group	*Characteristics*	*Class*	*Category*
Fusion welding	Weld metal is melted and parts joined without external pressure. The energy required for melting is supplied either by electric arc, gas torch, energy ray or by a chemical reaction.	Electric arc	Flux shielded: coated electrode, submerged arc, electroslag impregnated tape welding. Gas shielded: gas-metal-arc, gas-tungsten-arc, atomic hydrogen, plasma arc welding.
		Gas torch	Oxy-acetylene welding, pressure gas welding.
		Energy ray	Electron beam, laser welding.
		Chemical	Thermit welding.
Plastic welding	Weld metal is plastic but not melted. External pressure is applied to accomplish a weld.	Forge welding	Manual or machine forge welding.
		Pressure	Diffusion bonding, explosive welding, friction welding, ultrasonic, percussion welding.
		Electric resistance	Spot, seam, projection, upset, butt welding, radio frequency welding.
Brazing and soldering	Filler alloy is melted but not the weld metal. Brazing filler alloys have higher melting points than soldering alloys.	Brazing	Braze welding, brazing.
		Soldering	Silver soldering, soft soldering.

tion of welded metals may cause residual stresses and distortions. Heat treatment after welding can be used to relieve internal stresses, and weld jigs can prevent distortions. In designing welded components, weldments should not be located where much of the filler metal may be removed by later machining. Typical dimensional tolerances which may be held on average weldments are:

3 mm for small parts with little welding
6 mm for moderate sized parts with a small amount of welding
9 mm for large parts with a moderate amount of welding
9–12 mm for large parts with a large amount of welding

Mechanical fasteners include: 1, items that only provide temporary joints, e.g. nuts and bolts, lock washers and retaining rings; 2, items that provide permanent joints, e.g. rivets and metal stitching. Nuts and bolts are the most common means of joining materials. An ordinary nut can loosen if the forces of vibration overcome those of friction. Introducing a lock washer supplies an independent locking feature preventing the nut from loosening. For temporary fastening of metallic elements, machine screws, setscrews and cap screws are frequently used. Rivets are usually stronger than the thread type fastener and are more economical on a first cost basis. Riveting is the most common method of assembling aircraft and as many as 400 000 rivets can be used in one aeroplane. The materials usually used for making rivets are aluminium alloys, Monel metal, brass and steel. Mechanical joining methods have the disadvantage of the need to drill holes that lower the inherent strength of the material and could cause crevice corrosion. Metal stitching is used to join thin sections of metals and non-metals at high production rates. No precleaning, drilling, punching or hole alignment is necessary and this reduces the costs and increases production rates.

Adhesive bonding has reached a high level of development and reliability and is used widely in manufacturing and construction. The dollar value of adhesive bonded products per year rivals or exceeds the value of welded products. Adhesives have an advantage over other joining methods in that they can be applied to any surface, which enables them to join widely different materials, e.g. metal to glass, metal to metal, metal to polymer and polymer to polymer. The elimination of rivets and other fasteners results in smoother surfaces and considerable savings in weight and space. In addition, adhesive joints prevent fluid leakage and bimetallic corrosion at joints, as well as providing some degree of thermal and electrical insulation. Adhesives are available in solid, liquid, powder or tape form, which makes them suitable for mass production techniques. There are two types of adhesive action: 1, specific adhesion that results from interatomic or intermolecular action between the adhesive and the assembly; and 2, mechanical adhesion produced by the penetration of the adhesive into the assembly, after which the adhesive hardens and is anchored. Table 1.25 lists some common adhesives with their preparation properties. A serious disadvantage of adhesives is that, because most of them are organic materials, they cannot be used at high temperatures and their strength decreases rapidly as the temperature rises. Most adhesives are not stable above about 450 K, although some are usable up to about 550 K.

TABLE 1.25

CHARACTERISTICS OF SOME COMMON ADHESIVES

Adhesive	*Curing temperature (K)*	*Service temperature range (K)*	*Shear strength at room temperature* (MN/m^2)	*Shear strength at high temperature* (MN/m^2)
Butyral-phenolic	410–450	220–350	17·5	7 at 350 K
Room temperature epoxy	289–305	220–355	17·5	10·5 at 355 K
High temperature epoxy	365–450	220–450	17·5	10·5 at 450 K
Epoxy-nylon	395–450	22–355	42	14 at 355 K
Epoxy-phenolic	395–450	22–535	17·5	7 at 350 K
Neoprene-phenolic	410–450	220–355	14	7 at 355 K
Nitrile-phenolic	410–450	220–495	28	14 at 495 K
Polyimide	560–615	22–810	17·5	7 at 810 K
Urethane	297–495	22–350	17·5	7 at 350 K

1.9 MANUFACTURING BY MACHINING

After a component has been processed by casting, working or welding, it is usually finished by machining to improve its dimensional accuracy and surface finish. Usually machine tools are used for such operations, which can be divided into five basic metal cutting processes: drilling, turning, planing, milling and grinding. There are also some special metal removal processes such as chemical milling, electrochemical machining, ultrasonic machining, electric discharge and lasers. In the basic metal cutting processes a cutting tool is forced through the workpiece to remove excess material in the form of chips. The kind of surface produced by machining depends on the shape of the cutting tool and the path of the tool as it traverses through the material. Table 1.26 gives a list of the industrial machining processes and the expected dimensional tolerance and surface roughness under normal working conditions.

As machining is relatively expensive, it should not be performed unless it is necessary, and tolerances that are closer than necessary should not be specified. The economics of metal cutting can be improved by using high cutting speeds and tools with long lives. If the material to be cut gives discontinuous chips and needs less power for cutting, the economics are further improved. The ease with which one or more of the above factors can be realised for a given material is taken as a measure of its machinability. One of the methods for comparing machinability of materials is to determine the relative power required to cut them using single point tools. Another method is to use the machinability index, which is defined as follows:

$$\text{Machinability index \%} = \frac{\text{Cutting speed of material for 20 min tool life} \times 100}{\text{Cutting speed of SAE 1112 steel for 20 min tool life}}$$

In this definition the SAE 1112, or AISI B 1112 steel, is taken as the standard, and its machinability index is arbitrarily fixed at 100%. This steel contains some sulphur and is known as a free-machining steel (see Table 1.3). The machinability index of some common metallic materials is given in Tables 1.5 and 1.27. The higher the machinability index, the easier and the more economical it is to finish the material by metal cutting. The cutting tool material is also an important parameter in determining the cost of finishing. This will be discussed in detail in Chapter 14.

1.10 GENERAL CONSIDERATIONS IN SELECTING METALLIC MATERIALS

Discussions in this chapter have indicated the very wide range of properties and characteristics of metallic materials which makes the selection of the alloy that best fits the design criteria a very difficult task. The following considerations will help in making the selection process a little easier.

An important characteristic of metallic materials is that their elastic moduli are very difficult to change, e.g. it is impossible to stiffen a piece of

TABLE 1.26

CHARACTERISTICS OF INDUSTRIAL MACHINING OPERATIONS

Operation	*Machine tool most commonly used*	*Surface roughness* (μm)	*Minimum production tolerance* (μm)	*Remarks*
Turning	Lathe	0·8–15·0	± 0·025	Used for all surfaces of revolution. Single point tools.
Boring	Boring machine	0·4–8·0	± 0·002–± 0·03	Internal surfaces of revolution. Single point tools.
Shaping	Shaper	0·8–15·0	± 0·025	Flat surfaces or slots. Single point tools.
Planing	Planer	0·8–15·0	± 0·025	Flat surfaces or slots. Single tools.
Drilling	Drill	3·0–8·0	± 0·05	Cylindrical holes. Twin edge drills.
Milling	Milling machine	0·8–15·0	± 0·025	Flat and contoured surfaces and slots. Multipoint tools.
Broaching	Broaching machine	0·8–15·0	± 0·025	Contoured surfaces and slots. Multipoint tools.
Cylindrical grinding	Grinding machine	0·2–4·0	± 0·002	Cylindrical surfaces. Multipoint tools.
Surface grinding	Grinding machine	0·2–4·0	± 0·002	Flat and contoured surfaces and slots. Multipoint tools.
Chemical machining	Etching machine	0·2–4·0	± 0·075	Uses photo negative of template.

(*Contd.*)

TABLE 1.26 (*Contd.*)

Operation	*Machine tool most commonly used*	*Surface roughness* (μ*m*)	*Minimum production tolerance* (μ*m*)	*Remarks*
Electrochemical machining	Machine for ECM	0·2–4·0	± 0·05	For conducting surfaces. Workpiece is the anode in a plating process.
Electrodischarge machining	Machine for EDM	0·4–10·0	± 0·15–± 0·025	For conducting surfaces. Electrode negative of required form.
Ultrasonic machining	Ultrasonic machining machine	0·2–4·0	± 0·05–± 0·1	Faster than chipless machining operations. Liquid slurry carries abrasive powder.

TABLE 1.27

MACHINABILITY INDEX OF SOME COMMON METALLIC MATERIALS

Material	*Machinability index* (%)	*Hardness* (*BHN*)
Ferrous alloys		
Carbon steels and low alloy steels,	Table 1.5	
18–8 stainless steel	25	150–160
Tool steel (low W, Cr, C)	30	200–218
High speed steel	30	annealed
Grey cast iron (soft)	80	160–193
Grey cast iron (medium)	65	193–220
Grey cast iron (hard)	50	220–240
Malleable iron (standard)	120	110–145
Malleable iron (pearlitic)	80	200–240
Non-ferrous alloys		
Aluminium alloys	300–2000	35–150
Leaded brass	300–600	
Brass	200	
Bronze	150–500	55–210
Gun-metal	60	
Inconel	45	
Magnesium alloys	500–2000	50–75
Nickel alloys	20	
Zinc alloys	200	80–90

steel by heat treatment, and all steels regardless of composition or grade have elastic moduli in the relatively narrow range of 190×10^3 to 220×10^3 MN/m^2. To achieve a large change in the stiffness of a component, either its shape or the alloy base should be changed.

In almost all metallic materials, the strength increases and the ductility decreases with strain hardening and with decreasing temperature. Considerable changes in strength and ductility can be achieved by heat treatment, e.g. annealing increases ductility and softens the material, martensitic structures of steels are very hard but brittle, and precipitation of hard phases increases the strength. In hot worked products, the finishing temperature is very important in determining the properties of the material, e.g. lower finishing temperatures in hot rolled steels usually induce a fine grain and high toughness.

Wrought alloys are available in many size tolerances. Tolerances are relatively wider for hot worked products and this could cause difficulties in further processing, e.g. automatic machining where the work is held in collets. Cold worked products are usually supplied with narrower tolerances but residual stresses in them can cause unpredictable size changes during machining. Under these conditions stress relief treatments are necessary. The surface quality of wrought products is also important and a rough surface can cause difficulties in further processing, e.g. deep drawing of sheets or drawing of wires and tubes.

Differences in machinability can be a deciding factor when making the final selection among several candidate materials, especially in the cases where large amounts of metal must be removed by machining. Generally, steels with BHN less than 225 are machined easily and for average applications, the cost of machining rises rapidly above 250 BHN. Above 300 BHN considerable difficulties can occur in machining. It is preferable to avoid selecting materials with a machinability index of less than 40 if their processing involves considerable machining. Frequently the machinability can be improved by heat treatment, e.g. normalising low carbon steels to reduce hardness. Lead, phosphorus and sulphur have also been successfully added to different alloys to improve machinability, but at the expense of reduced strength and ductility.

The weldability of alloys is affected by their composition, e.g. plain carbon steels containing more than 0·3% C should be preheated and welded at a slow rate to reduce the cooling rate and avoid undesirable transformations.

Finally it should be mentioned that variations in design can have as much effect on component behaviour as variations in materials and heat treatment, e.g. stress concentrations, incorrect profiles or eccentric assemblies can obscure changes in material properties.

BIBLIOGRAPHY

1. R.A. Flinn and P.K. Trojan, *Engineering Materials and Their Properties*, Houghton Mifflin (Boston), 1975.
2. A. Guy, *Essentials of Materials Science*, McGraw-Hill (New York, Maidenhead England), 1976.
3. A. Guy, *Physical Metallurgy for Engineers*, Addison-Wesley (Reading Mass., London, Amsterdam), 1966.
4. Z.D. Jastrzebski, *The Nature and Properties of Engineering Materials*, 2nd Edn, John Wiley and Sons (New York), 1976.
5. *Materials Selector, Materials Engineering*, 1971.
6. *Metal Progress, Data Book*, American Soc. for Metals, 1976 and 1977.

7. *Metals Handbook,* Vol. 1, 'Properties and Selection of Metals', 8th Edn, American Soc. for Metals, 1961.
8. J.B. Moss, *Properties of Engineering Materials*, Butterworths (London), 1971.
9. E.C. Rollason, *Metallurgy for Engineers*, 4th Edn, Edward Arnold (London), 1973.
10. L.H. Van Vlack, *Materials Science for Engineers*, Addison-Wesley (Reading Mass., London, Amsterdam), 1970.

2

Polymeric Materials

2.1 INTRODUCTION

Polymeric materials encompass large-molecular-weight hydrocarbon materials such as plastics and rubber. Plastics can be defined as materials consisting mainly of combinations of carbon with oxygen, hydrogen, nitrogen and other organic or inorganic elements. Thermoplastics and thermosets are the two basic types of plastics; they differ in the degree of their intermolecular bonding. Thermoplastics have little or no cross bonding between molecules, and soften when heated and harden when cooled, no matter how often the process is repeated. Thermosets, on the other hand, have strong intermolecular bonding, which prevents the fully cured materials from softening when heated.

Rubbers are similar to plastics in that they are hydrocarbon polymeric materials with similar structure, and the difference is largely based on the degree of extensibility or stretching. Rubbers can be stretched to at least twice their original length and, upon release of the stress, return to approximately the original dimensions. Some grades of plastics approach this definition of rubber.

As a family, polymeric materials have some characteristics that distinguish them from other materials, particularly metals. Polymeric materials are essentially non-crystalline, non-conductors of heat and electricity and exhibit viscoelastic behaviour. With some exceptions, polymeric materials are resistant to chemicals and corrosive environments and have relatively low softening temperatures. Polymeric materials fall mainly into the category of commonly used materials, and the size of their production is similar to that of steel when compared on a volume basis. Some 3×10^5 separate polymers have already been synthesised and tested, and more are being developed.

Selection of a polymeric material for an engineering application is usually based on one or more of the following considerations:

1. Mechanical properties, such as tensile and compressive strengths, tensile modulus, elongation percentage, hardness, impact strength and heat deflection temperature.
2. Physical and chemical properties, such as specific gravity, coefficient of thermal expansion, dielectric constant, dielectric strength, dissipation factor and resistance to chemicals and corrosive environments.
3. Processing considerations, such as the ability to produce the required shape by the commonly used methods of compression moulding, transfer moulding, injection moulding or extrusion.
4. Cost and availability.
5. Sales appeal, such as colour and light weight.

The key advantage of plastics is the ability to produce accurate components, with excellent surface fiinish and attractive colours, at low cost and high speed.

2.2 POLYMERISATION MECHANISMS

The structure of a polymeric material is intimately related to the polymerisation mechanism involved in forming it. During polymerisation, a monomer, or other small molecule, attaches itself to a growing molecule in order to produce the polymeric molecule. In addition polymerisation, an initiator attracts one of the electrons of the carbon double bond of the monomer, and the other electron in turn attracts an electron from another monomer, and so the molecule grows. The rate of addition polymerisation can be very rapid, and is sometimes called chain-reaction polymerisation. A variety of chemicals including organic substances can be used as initiators and retarders to modify the reaction rate. Heat, light or pressure can also act as initiators to the addition reaction. Near the end of the reaction, diffusion of monomers to the ends of growing chains becomes progressively restricted, and the reaction slows down.

Addition polymers can be made to contain more than one kind of mer, e.g. vinylchloride and vinyl acetate. The resulting structure, called a copolymer, is comparable to a solid solution in metallic materials, and may have properties quite different from those of its components.

In contrast to addition reactions, which are primarily a summation of individual molecules into a polymer, condensation reactions form in steps between two or more different molecules which makes it a slower process. Furthermore, there is usually a by-product which must condense, hence the name of the reaction. This by-product is usually water or some other simple molecule, such as HCl or CH_3OH. Condensation reactions can produce either a chainlike molecule such as 6/6 nylon, or a network structure such as phenol–formaldehyde.

2.3 POLYMERIC STRUCTURES

The structure of a polymer affects its behaviour in a number of ways. Simple linear molecules are easy to move relative to each other, and the material is relatively weak and flexible, as in the case of polyethylene. On the other hand, if the linear molecules have projections on the side, their movement relative to each other becomes more restricted and the material is stronger, as in the case of polyvinyl chloride. These changes in molecule shape can be accomplished by changing the polymer composition or by changing the polymerisation mechanism. In all these cases the binding force between the different molecules is of the van der Waals type, which weakens with heating, leading to the melting of the material, i.e. it is thermoplastic. If the molecules crosslink however, a three dimensional network forms and heating merely helps to continue the bonding; the material is therefore thermosetting. Further raising of the temperature may cause the primary bonds of the chains, rather than the bonds between molecules, to break and degradation then takes place.

The above types of structure can be classified into two categories: 1, random arrangement of linear molecules; and 2, network structures consisting of one large molecule with an amorphous structure.

A third category is that of crystalline structures, which are important in the linear polymers because they lead to stiffer, stronger materials. Crystallinity usually reduces transparency because of the light scattering at grain boundaries. Crystallisation is seldom complete in polymers because only weak van der Waals forces are available for aligning the molecules, and a very large number of atoms must be manoeuvred into position. The tendency to crystallise is closely related to the structure and polarity of the molecule. Regular molecules with large side projections or branches show a stronger tendency to crystallise.

2.4 MECHANICAL BEHAVIOUR OF POLYMERIC MATERIALS

True thermosetting plastics such as phenolics and melamine have no yield point or plastic range, but have an elastic range extending essentially up to the fracture stress. Thermoplastics, on the other hand, in a normal environment continuously change deformation with increasing time when stressed, i.e. they exhibit creep. When the stress is removed, they recover the unstressed form to varying degrees over a period of time. Recovery is incomplete under conditions in which defects such as crazing or necking have become extensive.

When using polymers in applications where they have to bear stresses, the creep of the material must be assessed for the required time of stress application. Polymers cannot be regarded as linear elastic solids except in exceptional cases, and the non-linear nature of their viscoelasticity means that creep curves at several stress levels are necessary to define their performance, as shown in Fig. 2.1(a). This behaviour can be represented as stress versus time at constant strain, i.e. stress relaxation (Fig. 2.1(b)), or as stress versus strain at constant time (Fig. 2.1(c)). The curves of Fig. 2.1(c) indicate that the slope of the stress strain curves for polymeric materials is not constant as in metals but is time dependent, hence the name creep modulus. Creep modulus versus time curves are therefore obtained from the curves of Fig. 2.1(c) by calculating the slopes at constant strain as shown in Fig. 2.1(d).

The recovery of initial dimensions after load removal is dependent on the loading time, and is generally more rapid after short periods of time at low strains, than after long periods of time at high strains. The recovery time is an important parameter in applications involving intermittent loading.

It should be emphasised that the mechanical behaviour of polymers is very sensitive to temperature and in some cases to humidity. For some materials, the creep moduli can change by up to 4% for each 1 K change in temperature or each 1% change in relative humidity. This factor should be taken into consideration when comparing the mechanical properties of different polymers produced by different manufacturers.

2.5 PHYSICAL PROPERTIES OF POLYMERIC MATERIALS

The specific gravity of polymeric materials is a function of the weight per volume of the individual molecules and of the way they pack. The specific

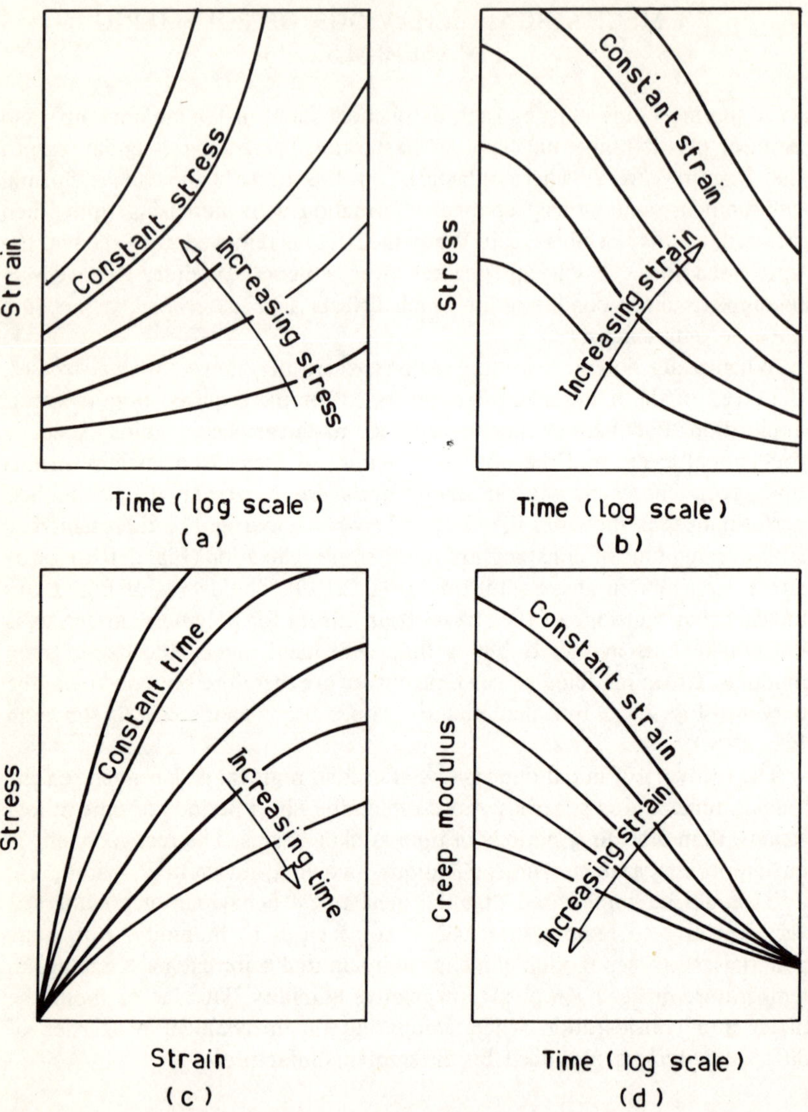

FIG. 2.1. Mechanical behaviour of polymeric materials. (a) Creep curves at constant stress, (b) stress relaxation curves at constant strain, (c) stress–strain curves at different testing speeds, (d) variation of creep modulus with strain and time.

gravity of simple hydrocarbon polymers like polyethylene and polystyrene is 0·9 to 1·1. Substitution of a relatively heavy atom such as chlorine or fluorine for hydrogen gives polyvinylchloride, with a specific gravity of 1·2 to 1·55, and Teflon, with one of 2·1 to 2·2. The crystalline form of a polymer is always denser than the amorphous form, due to more efficient packing.

The coefficient of expansion of polymers is approximately 2 to 20 times as great as that of steel, and is not usually linear. This means that if a polymer is to replace a metal in a component with close tolerances, due allowance should be made. Fillers may be incorporated to reduce the coefficient of expansion, but may, if of fibrous form, give anisotropic characteristics.

Polymers are non-conductors of electricity and some are among the best insulators available; therefore, except in specialised applications, it is likely that the electrical requirements of a design will be met by the dimensions needed for mechanical functioning. As with the mechanical properties, electrical characteristics vary with time and temperature and sometimes with humidity. The frequency and the direction of the electric field, as well as the direction of applied mechanical stresses in relation to molecular orientations, can affect the measured values of electrical properties. Treatments that partially destroy the polymeric structure, such as gamma ray irradiation or thermal degradation, can expose irregularities that contain donor and acceptor sites and these give some conductivity. Adding radicals along the polymer chain can also result in some conductivity.

Generally, the index of refraction of a polymeric material is proportional to its density. If the densities of the amorphous and the crystalline forms of the polymer are close, there is little dispersion as the light passes through a mixture of the forms, and the material is transparent. On the other hand, if there is a substantial difference between densities, the material is opaque. Also, crystalline materials can be transparent if they are cooled rapidly so that the crystallites are very small and shorter than the wavelength of the light to be transmitted. Fillers such as asbestos and carbon black make the material opaque.

2.6 THERMOPLASTICS

Of the two basic types of plastics, the thermoplastics family is the larger, both in number and in volume used. Generally, thermoplastics soften into a plastic state of high viscosity when heated for processing. Most thermoplastics are furnished chemically complete, polymerised, and in the form of small pellets or cubes ready to mould. Some are also available as liquid resins for casting, laminating or coating. Thermoplastics are usually used in the unfilled form.

Polyolefins are based on the ethylene monomer and are the most widely used group of plastics. Polyethylenes and polypropylenes are the main members of this group. Polyethylenes are relatively cheap, tough, resistant to chemicals and have excellent dielectric strength. Because of their low water absorption and permeability, polyethylenes are widely used in sheet form as moisture barriers. The different types of polyethylenes can be classified according to their density into low density, medium density and high density. Table 2.1 lists some representative properties and uses of polyethylenes. Polypropylenes are similar in many respects to high density polyethylenes and can be classified into three basic groups: homopolymers, copolymers and reinforced (Table 2.1). Copolymers are produced by adding other types of olefin monomers to improve properties like low temperature toughness. The mechanical properties can be further improved by adding fillers or fibres of glass or asbestos. The ability of polypropylenes to withstand fatigue loading has made them popular for surgical implant hinges. Although polypropylenes have low creep resistance, reinforcing them with asbestos can raise their maximum service temperature to about 400 K.

Polystyrenes are second only to polyethylenes in volume of use. They are based on the styrene monomer with low crystallinity and considerable branching. Various modifications create a wide range of structures and properties. Polystyrenes are generally rigid, low cost and have exceptional insulating properties. Although they have reasonable tensile strength, they are subject to creep and their maximum service temperature is limited to about 350 K. The impact grades and glass filled types are used widely for engineering parts, and polystyrene foams are highest in volume use of all the plastics. Table 2.1 lists the properties of some commonly used polystyrenes.

Vinyls are the third most widely used plastic. Like polyethylenes and polystyrenes, they are versatile and low in cost. Vinyls have low crystallinity and branched molecules. Modifications can change the vinyls from a stiff and rigid state to a flexible and rubber-like one. Vinyls have excellent chemical resistance, high abrasion resistance and good electrical resistivity. Their maximum useful temperature is about 350 K. The most widely used vinyls are polyvinyl chloride (PVC) and PVC–acetate copolymer. Table 2.1 lists the properties of some types of PVC.

Acrylic plastics are mostly based on methyl methacrylate polymers,

TABLE 2.1

SOME PROPERTIES AND USES OF SELECTED THERMOPLASTICS

Material	*Tensile strength* (MN/m^2)	*Tensile modulus* (MN/m^2)	*Elongation* (%)	*Hardness*	*Izod impact* (*joule*)	*Expansion coefficient* ($\times 10^5 m/m/K$)	*Heat deflection temperature** (*K*)	*Specific gravity*	*Relative cost†*	*Typical applications*
Polyethylene										
Low density	7–21	140	50–800	Rr10	27	19·8	309	0·92	1·0	High frequency insulators, pipes for chemicals, battery parts, blow-moulded bottles and containers, flexible-film packaging.
Medium density	14–21	280	50–800	Rr14	2·7–22	16·2	317·4	0·93	1·0	
High density	21–35	700–1 400	1 030	Rr65	1·36–6·8	13·5	322	0·96	1·0	
Polypropylene										
Homopolymers	35	1 200	150	Rr90	0·54–2·0	8·6	330	0·91	0·8	Film and fibre, housewares, appliances.
High impact copolymers	27	—	400	Rr65	5·4–8·1	—	323	0·9	1·1	Luggage, seating, housings.
Talc filled, 40%	35	3 600	5·0	Rr95	0·54	—	354	1·23	1·0	Covers, cases, containers, mechanical parts.

(*Contd.*)

TABLE 2.1 (*Contd.*)

Material	*Tensile strength* (MN/m^2)	*Tensile modulus* (MN/m^2)	*Elongation* (%)	*Hardness*	*Izod impact* (*joule*)	*Expansion coefficient* ($\times 10^5 m/m/K$)	*Heat deflection temperature** (*K*)	*Specific gravity*	*Relative cost†*	*Typical applications*
Polystyrene										
General purpose	35–56	3 300	1·5–4·0	Rm74	0·27–0·54	7·4	340–373	1·05	1·0	Thin parts, housewares, foam products.
High impact	23–32	2 200	25–60	Rm60	1·36–3·8	7·0	340–370	1·04	1·0	Containers, appliance housings.
30% glass fibre	77–100	8 800	—	Rm90	3·8	3·2	370	1·29	1·75	Automobile dashboard skeletons, frames, fan blades.
Polyvinyl chloride										
General purpose	7–28	20	400	Sa75	—	14·4	—	1·40	1·6	Appliances, grips, sheet for chemical tanks.
Flexible PVC	14–21	15	250	Sa85	—	14·4	—	1·35	1·6	Communication and low-tension cables.
Rigid PVC	35–63	2 000–4 200	100	Rr115	1·4–27	9–11	345	1·40	1·3	Fume hoods and ducts, tanks, piping.

Acrylic (MMA)										
Cast, general purpose	42–84	2 800	—	Rm91	0·68	5·4–7·2	372	1·19	1·4	Transparent enclosures, radio and TV parts, fixtures.
Moulding grade	63–77	2 800	—	Rm95	0·5	5·4–7·2	361	1·18	1·7	Reflectors, protective goggle lenses.
High impact	42–63	2 100	25–40	Rm45	1·4–5·4	7·2–11	350	1·11	2·2	Control knobs, pump parts, tool handles.
Nylons										
Nylon 6/6	84	3 300	60–300	Rr118	1·36–2·7	8·1	377	1·14	3·2	Bearings, gears, bushings, brush blocks, tubing.
Nylon 6/12	62	21 000	150–340	Rr114	1·36–2·7	9	355	1·07	5·8	Wire jacketing, pump housings, bearings, gears.
Acetal										
Homopolymer	70	3 700	25–75	Rm94	1·9–3·1	9–14·4	397	1·42	2·5	Appliance parts, gears, bushings, aerosol bottles, plumbing, consumer products.
Homopolymer 22% TFE fibre	53	2 800	12–21	Rm78	0·95–2·3	9–14·4	373	1·52	15	
Copolymer	62	2 900	60–75	Rm80	1·63–3·5	8·46	383	1·41	1·9	
Polycarbonate										
Unfilled	63	2 400	110	Rm70	16–22	6·75	405	1·2	3·6	Electrical and electronic components, lenses, helmets, tool housings.

(*Contd.*)

TABLE 2.1 (*Contd.*)

Material	*Tensile strength* (MN/m^2)	*Tensile modulus* (MN/m^2)	*Elongation* (%)	*Hardness*	*Izod impact* (*joule*)	*Expansion coefficient* ($\times 10^5 m/m/K$)	*Heat deflection temperature** (*K*)	*Specific gravity*	*Relative cost*†	*Typical applications*
ABS										
Medium impact	46	2 500	6–14	Rr111	5·4	8·46	367	1·05	1·4	Pipes and fittings, appliance housings, chromium-plated parts, fume hoods and ducts, telephones, refrigerator components.
High impact	42	2 300	10–35	Rr103	8·8	9·5	372	1·04	1·6	
Very high impact	33·6	1 750	15–50	Rr88	10·8	11·0	369	1·02	1·7	
Heat resistant	50·4	2 450	5–20	Rr111	3·12	6·7	387	1·05	1·8	
Fluoroplastics										
PCTFE	28–42	1 400	160	Sd76	4	4·5	—	2·1	20	Chemical pipes, pump parts, cables.
PTFE	14–49	700	100–450	Sd58	6·1	9·9	—	2·16	10	Chemical pipes, valves, pump parts, electrical equipment.
High temperature plastics										
Polyimide unfilled	91	3 150	7–9	Rm97	1·36	5·0	633	1·43	—	Bearings, seals, gears, high temperature mechanical parts.
Polyimide 40% graphite	53	5 300	2–3	Rm73	—	3·2	633	1·65	—	
Polysulphone	71·4	2 500	50–100	Rr120	1·63	5·6	447	1·24	6·3	

*Heat deflection temperature is taken for a stress of 1·8 kN/m^2 (264 psi).
†Relative cost is calculated by comparing the material cost with the cost of low-density polyethylene, 1976 prices were used in

modified by copolymerisation or blending with other monomers. Acrylics are strong and stiff, but regular grades are brittle. High impact grades are produced by blending with rubber stock. Acrylics are available as cast sheets, rods, tubes or blocks, and as powders for injection moulding and extrusion. Table 2.1 lists the properties of some acrylic plastics.

Polyamides (Nylon) have basically linear molecular structures with a relatively high degree of crystallinity, which gives them good mechanical properties. They also have a very high abrasion resistance and good frictional characteristics. Their major disadvantage is their moisture absorption and resulting dimensional changes. Nylon 6/6 is the most widely used of the Nylons and types 6/10, 6/11 and 6/12 are less moisture absorbent than other types. Nylons are processed mainly by moulding and extruding, but small parts can be processed by powder sintering. Table 2.1 lists the properties and uses of some Nylons.

Acetals are highly crystalline and among the strongest and stiffest thermoplastics. Their excellent creep resistance and low moisture absorption give them excellent dimensional stability. They are useful for continuous service up to about 380 K. Acetals are processed mainly by moulding or extruding but some parts are blow moulded. Table 2.1 lists some properties of acetal plastics.

Polycarbonate is a linear, low-crystalline, transparent plastic. It is the toughest and one of the hardest plastics. Polycarbonate also has excellent electrical resistivity and its maximum service temperature is about 410 K. Although it has negligible moisture absorption, it is easily attacked by solvents. Polycarbonate has very low and uniform mould shrinkage, and can be easily processed by extrusion, injection moulding, blow moulding, and vacuum forming. Table 2.1 lists some properties of polycarbonate plastics.

The ABS family of plastics is composed of monomers of acrylonitrile, butadiene, and styrene. They have a good balance of properties which they maintain over the range of temperatures 230 to 380 K. They are available in medium, high and very high impact grades, as well as heat resistant, transparent, flame retardant and expanded grades. ABS plastics are readily processed by extrusion, injection moulding, blow moulding, calendering, and vacuum forming. ABS plated components are in wide use as a replacement for metals, e.g. in automobiles and appliances. Table 2.1 lists some properties of selected ABS grades.

Fluoroplastics are composed basically of linear molecules with fluorine replacing some or all of the hydrogen atoms. They are highly crystalline with high molecular weight. As a class fluoroplastics rank among the best plastics for chemical resistance and for high temperature performance up to 530 K. Polytetrafluoroethylene (PTFE) is the most widely used fluoroplastic because of its high service temperature. PTFE cannot exist in a true molten state and cannot be formed by moulding. It is usually fabricated by compacting the resin and then sintering. Table 2.1 lists some properties of selected fluoroplastics.

Several groups of thermoplastics have been developed to withstand temperatures in excess of 475 K for extended periods. These include polyimide, polysulphone, polyphenylene sulphide and polyarylsulphone. In addition to their high temperature resistance, these plastics have high strength and tensile modulus, and excellent resistance to solvents. Their major disadvantage is difficulty in processing them. Their high price limits their use to specialised applications. Table 2.1 lists some properties of selected high temperature plastics.

2.7 THERMOSETTING PLASTICS

The presence of the strong covalent bonds between molecules in thermosetting plastics makes them harder and more brittle but gives them more thermal stability and creep resistance than thermoplastics. Thermosetting plastics are generally more difficult to process than thermoplastics but once cured will not return to their original state. Before curing, thermosetting plastics are composed of a resin system and fillers. Sometimes reinforcements are also added. Thermosetting plastics are usually classified according to the resin component, which consists of a polymer, curing agents, hardeners, inhibitors and plasticisers. The resin component influences the dimensional stability, heat and chemical resistance, electrical properties and flammability. The fillers usually have a major effect on mechanical properties.

Phenolics (phenol formaldehyde) are relatively cheap and are readily moulded with good stiffness and impact resistance. Phenolics can be classified as: general purpose with wood flour fillers, shock resistant with paper or fabric fillers, heat resistant with mineral or glass fillers, electrical grade with mineral fillers, chemical grade with no fillers, and rubber phenolics. Glass fibres are added in some cases to improve the mechanical behaviour, as will be discussed in Chapter 5. Phenolics can be processed by compression, transfer or injection moulding, as well as by extrusion. Table 2.2 lists some properties of selected phenolics.

Epoxies are one of the most versatile plastics and are used in high

TABLE 2.2

SOME PROPERTIES AND USES OF SELECTED THERMOSETTING PLASTICS

Material	*Tensile strength* (MN/m^2)	*Tensile modulus* (MN/m^2)	*Elongation* (%)	*Hardness*	*Izod impact (joule)*	*Expansion coefficient* ($\times 10^5 m/m/K$)	*Heat deflection temperature** (*K*)	*Specific gravity*	*Relative cost†*	*Typical applications*
Phenolic										
General purpose	35–63	800	—	Re95	0·44	3·8	447	1·38	—	Mechanical applications (pulleys, wheels)
Shock and heat	28–63	14 000	—	Re85	2·17	2·3	427	1·83	—	Motor housings, handles.
Heat (mineral)	35–49	11 000	—	Re85	0·54	1·8	450	1·53	—	Appliance connector plugs, handles.
Electrical (mineral)	28–56	6 000	—	Re87	0·41	3·96	427–477	1·52–1·67	—	Coil forms, ignition parts, condenser housings.
Epoxy										
Cast rigid	63–105	3 200	—	Rm106	0·5	5·6	439	1·20	1·3	Encapsulation of electronic components, tools and dies.
Moulded	56–140	14 000	—	B78	1·3–5·5	3·6	464	1·91	1·3	Moulded coils, relay assemblies.
High strength laminate	350–490	32 000	—	B71	13·6–41	3·6	—	1·84	1·3	Tools for metal forming, aircraft parts.

(*Contd.*)

TABLE 2.2 (*Contd.*)

Material	*Tensile strength* (MN/m^2)	*Tensile modulus* (MN/m^2)	*Elongation* (%)	*Hardness*	*Izod impact* (*joule*)	*Expansion coefficient* ($\times 10^5 m/m/K$)	*Heat deflection temperature** (*K*)	*Specific gravity*	*Relative cost†*	*Typical applications*
Polyester										
Unfilled	56	2 400	200-300	Rm117	1·63	11·3	327	1·31	2·7	Gears, bushings, bearings, switch housings.
30% glass fibre	123	7 700	3	Rm90	2·3	2·3–9·7	486	1·54	2·9	Pumps, automotive parts, connectors.
Alkyd										
Granular (mineral)	42–63	16 000	—	Re85	0·42	2·9	464–419	2·2	1·4	Tube bases and sockets, potentiometers.
Silicone										
30% glass fibre	42	17 500	—	Rm90	4–20·3	5·76	755	1·88	—	Premium plastics used in critical or high performance applications, e.g. aircraft, aerospace, and electronics.
Silica reinforced	28	11 000	—	Rm82	0·41	4–8	755	1·93	—	

*Heat deflection temperature is taken for a stress of 1·8 kN/m^2 (264 psi).
†Relative cost is calculated by comparing the material cost with the cost of low-density polyethylene, 1976 prices were used in the calculations.

performance applications where their high cost is justified. They are available in a wide variety of forms, both liquid and solid, and are cured into the finished product by catalysts. The epoxy resins give good chemical resistance, excellent bonding properties, and exceptionally high strength, especially when reinforced with glass fibre or other similar fibres. Liquid epoxies are processed by casting, and powders by moulding. Typical uses and properties of selected epoxies are given in Table 2.2.

Polyester resins are copolymers of a polyester and, usually, styrene. Unreinforced polyesters have only limited use, and the majority of products are glass reinforced either as mouldings or laminates. Alkyd plastics are composed of polyester resin and, usually, a diallyl monomer with inorganic fillers. Their low moisture absorption and good dielectric strength makes them particularly suitable for electrical applications. Table 2.2 lists some properties and uses of polyester and alkyd plastics.

Silicones are semi-organic polymers composed of monomers in which oxygen atoms are attached to silicon atoms together with radicals. Silicones are available as liquids, elastomers or rigid solids depending on the type of radicals present. Silicone moulding compounds consist of silicone resin, inorganic filler and catalyst. They have excellent resistance to heat, water and certain chemicals. Typical uses and properties are given in Table 2.2.

2.8 RUBBERS

Rubbers are hydrocarbon polymers with sulphur or other additives cross-linking the polymers to each other, producing a thermoset-like structure. The rubber becomes stronger but less extensible as the amount of cross-linking between sulphur and carbon atoms increases. When rubber is stretched, the polymer chains tend to straighten and become aligned which increases crystallinity and strength. Table 2.3 lists some properties and applications of selected rubbers which will be discussed below.

Natural rubber is a homopolymer of the isoprene monomer. A number of grades of rubber are made from the raw material by adding fillers like carbon black, silica and silicates, as will be discussed in Chapter 5. The soft grades have excellent resilience and good abrasion resistance, with low hysteresis and heat build-up under repeated loading. Styrene Butadiene Rubbers (SBR) are copolymers of butadiene and styrene and are similar in many ways to natural rubbers. They are the most widely used group of rubbers because of their low cost and their use in automobile tyres. A wide range of properties can be obtained by changing the butadiene to styrene

TABLE 2.3

SOME PROPERTIES AND APPLICATIONS OF SELECTED RUBBERS

Material	*Tensile strength* (MN/m^2)	*Elongation* (%)	*Hardness shore A*	*Specific gravity*	*Typical applications*
Natural rubber	28	700	30–90	0·92	Conveyor belts, tyre products, sound damping, gaskets.
Styrene butadiene	24·5	600	40–90	0·94	Auto tyres, conveyor belts, gaskets, sound damping.
Neoprene	28	600	30–90	1·24	Heavy duty conveyor belts, V-belts, footwear, brake diaphragms, motor mounts, gaskets.
Butyl	21	800	40–80	0·92	Cable insulation, encapsulating compounds, coated fabrics, high pressure steam hoses, machine mounts.
Silicone	8·4	700	30–85	0·98	Seals, gaskets, 0-rings, insulation for wire and cable and for encapsulation of electronic components.
Fluorocarbon	17·5	300	60–90	1·85	Brake seals, 0-rings, diaphragms.
Hypalon	21	500	50–90	1·18	High temperature conveyor belts, foot-wear, seals, gaskets, spark plug boots.

ratio, and those grades containing more than 50% styrene are usually considered as plastics. SBRs have excellent impact and abrasion resistance but poor chemical resistance.

Neoprene (chloroprene) is a synthetic rubber which is chemically and structurally similar to natural rubber. It has excellent resistance to oils, chemicals, sunlight, weathering, ageing and ozone, and retains its properties at temperatures up to 390 K. In addition, it has excellent resistance to permeation by gases. Butyl rubbers consist of copolymers of isobutylene and a few percent of isoprene. They are similar in many ways to natural rubber and are one of the lowest priced synthetics. Because of their excellent dielectric strength, they are widely used for cable insulation and similar electrical application.

Some silicones can be considered as rubbers in view of their extensibility. Silicone rubbers are the most stable of all rubbers with excellent resistance to high and low temperatures, oils and chemicals. They maintain their electrical properties over the range of temperatures 200 to 540 K. All silicone rubbers can be classified as high performance, high price materials.

2.9 PROCESSING OF POLYMERIC MATERIALS

Two of the main reasons for the fast expansion of the industrial use of polymers are: 1, their versatility; and 2, the ease with which they can be manufactured into complicated shapes, in one step, with very little need for further processing or surface treatment. The common production methods for processing polymeric materials are: compression moulding, transfer moulding, injection moulding, casting, extrusion and sheet forming. Composites are usually produced by laminating or filament winding, and foams are produced by foaming processes. Polymers can also be joined and sometimes machined. Each of the processing methods has certain advantages and limitations that have a direct bearing on component design, material selection and final cost. Not every polymer is suitable for each process, and selection should aim for the optimal material–process combination for the given design and application. Table 2.4 gives a comparison of common plastics forming processes.

Compression moulding involves feeding the proper amount of material into the mould, and then squeezing it with a ram to fill the cavity. The mould is kept closed long enough to allow the formed material to harden. This process is usually used for thermosetting plastics and the heat required for curing is supplied through the mould walls. When the material is supplied as loose powder or pellets, several minutes are needed for each cycle. The cycle time can be reduced by as much as 50% if the material is cold compressed in the shape of a preform. Another time saving process is to cold press the material to the required form, and then bake it in an oven to cure it. This latter process is similar to the powder metallurgy techniques described in Chapter 1. Compression moulding is most suitable for simple shapes with uniform wall thicknesses, preferably not more than 3 mm.

In transfer moulding, the material is first heated and compressed in a chamber, and then forced through an orifice into the mould cavity. This

TABLE 2.4

COMPARISON OF COMMON PLASTIC FORMING PROCESSES

Process	*Possible shapes*	*Possible materials*	*Surface finish*	*Metal inserts*	*Dimensional accuracy*	*Mould cost*	*Production rate*
Casting	Sheets, rods, tubes, simple shapes	Liquids	Very good	Yes	Poor	Very low	Low
Compression moulding	Intricate shapes	Wide choice	Very good	Yes	Good	High	Medium
Injection moulding	Intricate shapes	Thermoplastics	Very good	Yes	Very good	Very high	High
Transfer moulding	Intricate shapes	Wide choice	Very good	Yes	Very good	High	Medium
Extrusion	Rods, tubes, sheets, long shapes	Thermoplastics	Good	No	Fair	Low	High
Lamination	Sheets, rods, tubes	Fair choice	Good	Possible	Good	High	Low
Vacuum forming of sheets	Shapes from sheets	Thermoplastics	Fair	Yes	Fair	Low	Low

process involves a more expensive mould than compression moulding, but gives closer tolerances and more uniform density in the product.

Injection moulding is used to produce more thermoplastic products than any other process. Raw material is fed by gravity into a pressure chamber ahead of a plunger. As the plunger advances, the polymer is forced into a heating chamber and then through a nozzle to the mould cavity. In this process the mould remains cool so that the material solidifies as soon as the mould is filled, and the complete cycle requires only a few seconds. Intricate parts can be made to close tolerances with no need for second operations, but the mould can be so expensive that the production of as many as 10 000 components may be required for economical operation.

Casting techniques are employed for resins that are available in liquid form, such as phenolics, polyesters, epoxies and acrylics. Thermoplastics that can be softened enough to pour, such as some celluloses, can also be cast. A catalyst is usually added to unpolymerised resins and, after being poured into the mould cavity, they are heated to cure them. The mould material is usually lead, rubber, glass or plaster, and it is formed around a model of the required shape. The process is inexpensive but slow.

Extrusion involves feeding the material from a hopper into a screw chamber where it is preheated, compressed and then forced through a heated die onto a conveyor belt. The extrusion process is suitable for the production of sheets, tubes and shapes with re-entrant angles. Thermoplastics, thermosets and rubbers can be extruded.

Sheet forming, or vacuum forming, is used for thermoplastic sheets where they are softened by heating and then formed into a thin-walled component by suction or blowing against the mould. The unit costs for sheet forming, for both material and labour, are relatively high, and it is justified only for large components.

Laminated plastics are made by pressing and heating impregnated reinforcing materials, either under high pressure or under low pressure. The resins ordinarily used for impregnation are phenolics, melamines, silicones and epoxies. The reinforcing material is usually glass, in the form of fabric or fibres, but asbestos, boron, cotton or nylon fibres are also used. This method is usually used for thermosetting plastics and for some thermoplastics. In the filament winding method, continuous strands of glass roving or other filament materials are wound on a mandrel and impregnated, during or after winding, with resin. After the resin sets, the mandrel is removed. Filament winding is used to produce tanks and pressure vessels, rocket tubes, motor cases, pipes, etc. This process is expensive and is only used when a high strength/weight ratio is required.

Foams are made by chemically or mechanically expanding a polymer. Chemical methods involve adding a low-boiling-point solvent or chemicals and heating to release vapour or gas, or including chemicals with the components of the polymer, to form gas when mixed. Mechanical methods include: injection of a gas under pressure into the soft polymer; mechanical aeration or frothing by mixing; and mixing fine metal particles and gas with a polymer mass under pressure, causing bubbles to form on the metal particles when the pressure is released.

Joining of polymers can be achieved by cementing with solvents, adhesive bonding, thermal welding and mechanical fastening. Cementing is applicable to soluble thermoplastics like acrylics, polystyrenes and some vinyls. The strength of the joint is comparable to that of the parent material. Adhesive bonding is suitable for all polymers and was described in Section 1.8. Thermal welding can be applied to most thermoplastics, where they are softened by heating and pressed together to form the joint. The source of heat can be hot gas, a heated tool, induction heating or friction. Joining of polymers by mechanical fastening is similar to joining of metals, as discussed in Section 1.8. Machining of polymers is also similar to machining of metals, Section 1.9, but the following differences should be kept in mind. Polymers do not dissipate the heat generated in cutting, and thermosets may deteriorate or char, while thermoplastics may soften and clog cutters if overheated by friction with the cutting tool. This can be overcome by using sharp cutting edges and smooth, polished tool faces. Coolants are also commonly used in cutting polymers. Most plastics can be cut at high speeds, and cutting conditions similar to those used for brasses are usually adopted.

While the finishing costs of a metallic component can constitute a considerable proportion of the total cost, the majority of plastic products require very little finishing or decorative treatment. In some cases fins and rough spots can be removed, and smoothing and polishing done, by barrel tumbling with suitable abrasives or polishing agents. Plastic parts can be electroplated after coating with some electrically conductive material to which the plating material will adhere. This is accomplished by chemical means and is referred to as 'electroless metal plating'. The most commonly electroplated polymers are ABS, polysulphone and polypropylene. Coating of polymers with aluminium is usually achieved by metal vapour coating techniques.

2.10 DESIGN CONSIDERATIONS FOR POLYMER COMPONENTS

In every casting process in which a fluid or semifluid material is introduced into a mould cavity and permitted to solidify into a desired shape, the following basic considerations should be kept in mind. Many of the parameters in metal casting and moulding apply to polymer moulding. Taper is essential to remove parts from moulds, and a taper of 1° to 3° is usual for polymers. Thick sections should be avoided as they retard the moulding cycle and require more material. Ribs, beads and flanges should be used instead. However, flow lines or dimples may appear opposite ribbed surfaces if they are not proportioned correctly. Fillets must be added to facilitate the flow of material into the mould cavity, to reduce stress concentrations, and to lessen warping after moulding. Wall thicknesses should not be less than 2 mm for most polymers.

Inserts can cause difficulties in assembly, especially when clips or threads are used. When it is necessary to use inserts, they should be surrounded by enough material to avoid having the insert break loose in service.

Adequate dimensional tolerance makes economical production possible. A minimum tolerance of $\pm 0{\cdot}05$ mm should be allowed in the direction parallel to the parting line of the mould. In the direction at right angles to the parting line a minimum tolerance of $\pm 0{\cdot}2$ mm is desirable. Larger tolerances should be allowed in filled polymers. Large, plain, flat surfaces should be avoided on polymer components in view of their lack of rigidity. Either ribbing or doming should be employed instead.

2.11 GENERAL CONSIDERATIONS IN SELECTING POLYMERIC MATERIALS

Discussions in this chapter have indicated the wide variety and range of behaviour of available polymeric materials. However, there are a number of factors that are common to almost all polymers, and that should be kept in mind when selecting polymeric materials for a given application. Polymers are organic synthetic materials and are subject to the kind of influences typical of organic materials, such as the effect of high temperature, water, ultraviolet light, chemical attack and so on. These effects may oxidise, hydrolyse or cause depolymerisation and degradation of the material. The life of a polymer in service is an important parameter, and depends on the material composition, treatment and the intensity of the environmental attack. Polymers can be divided roughly into two categories. The high volume, reasonably priced materials like polyethylene, polystyrene and polypropylene are at one end of the scale while the high performance, high price materials like silicones, polycarbonates and fluoroplastics are at the other end. Glass reinforced and otherwise modified grades fall between the two ends of the scale.

Polymers should not simply be substituted for other materials, instead the design should make use of the advantages of polymers and should overcome their limitations. Polymers are viscoelastic materials and their properties are time dependent and very sensitive to temperature variations. A vast amount of information on polymeric materials is available from manufacturers in the form of publications; no two materials are exactly the same, and the variations within one material can be numerous. Direct comparisons are difficult in many cases because the materials are not tested according to the same standards. This is especially the case for properties like fire resistance, electrical behaviour and stress cracking resistance. The main standards used by the different manufacturers are: ASTM, American Society for Testing of Materials; BS, British Standards; ISO, International Standards Organisation; and DIN, Deutsche Industrie Norm.

BIBLIOGRAPHY

1. F.Z. BILLMEYER, *Textbook of Polymer Science*, Wiley-Interscience (New York), 1962.
2. R.A. FLINN and P.K. TROJAN, *Engineering Materials and Their Applications*, Houghton Mifflin (Boston), 1975.
3. Z.D. JASTRZEBSKI, *The Nature and Properties of Engineering Materials*, 2nd Edn, John Wiley and Sons (New York), 1976.
4. H. LEWIS, *Engineering Materials and Design*, September 1974, 17.
5. J.B. MOSS, *Properties of Engineering Materials,* Butterworths (London), 1971.
6. H.W. POLLACK, *Materials Science and Metallurgy,* Reston Pub. Co., Prentice-Hall (New Jersey), 1973.
7. L.H. VAN VLACK, *Materials Science for Engineers,* Addison-Wesley (Reading, Mass., London, Amsterdam), 1970.
8. C.C. WINDING and G.D. HIATT, *Polymeric Materials,* McGraw-Hill (New York, Maidenhead, England), 1961.

3

Ceramic Materials

3.1 INTRODUCTION

Ceramic materials are combinations of one or more metals with a non-metallic element, which is usually oxygen, but can be carbon, nitrogen, or boron and carbon. This definition of ceramics covers a wide variety of materials, with different properties because of the many possible combinations of metallic and non-metallic atoms. Examples of ceramic materials are brick, stone, concrete, abrasives, porcelain enamels, glasses and refractory compounds like oxides, carbides, nitrides, silicides and borides. Brick, stone and concrete will be discussed in Chapter 4 as they are used mostly as materials of construction. Glasses are non-crystalline, and this distinguishes them from the other ceramics which are normally crystalline. Crystalline ceramics have high stability, and on the average have higher melting points and greater chemical resistance than metals and organic materials. They are generally the hardest of the engineering materials, but are extremely brittle. Like organic materials, ceramics are usually insulators to heat and electricity due to the absence of free electrons in them. However, at elevated temperatures, with more thermal energy, they conduct electricity, but not as well as do metals.

The American Ceramic Society has classified ceramics into the following groups: whitewares, glass, refractories, structural clay products and enamels. Of these, whitewares, refractories, and glass will be discussed here while structural clay products will be discussed in Chapter 4. The use of enamels for corrosion protection will be discussed in Chapter 6.

3.2 STRUCTURE OF CERAMIC MATERIALS

The crystal structures of ceramics are complex, since they have to accommodate more than one element. Generally, crystal structures of ceramics can be grouped into two main categories, close packed structures and silicate structures. In close packed structures a relatively efficient packing is developed when smaller atoms are placed in sites that would not have been occupied if all the atoms were of the same size. In silicate structures the primary structural unit is the SiO_4 tetrahedron, in which one silicon atom fits interstitially among four oxygen atoms. The interatomic forces generally alternate between ionic and covalent bonds, which leave relatively few free electrons in ceramic structures as compared with metals. The strong ionic and covalent bonds give the ceramics their high hardness, stiffness and stability.

Vitreous silica has a three dimensional framework structure, similar to crystalline silica in its short range arrangement, but does not have the long range repetitious crystalline pattern. The non-crystalline structure of glass can readily adjust itself to the presence of other atoms by varying the number of ionic to covalent bonds.

At the macrostructural level, ceramic materials can have one of the following types of structures:

1. Crystalline bodies cemented together by a vitreous or glassy matrix, the glassy phase usually being the weaker of the two phases.
2. Crystalline bodies, or holocrystalline bodies, where the glassy phase is eliminated and the resulting material is stronger.
3. Non-crystalline structures or glasses.

In manufactured ceramic components a holocrystalline body can be covered with a thin glassy ceramic coating to make it impervious to moisture, and to provide special surface properties.

3.3 MECHANICAL BEHAVIOUR OF CERAMIC MATERIALS

Generally, the properties of ceramic materials, like those of other materials, depend on their structure. Variations in mechanical behaviour result from the various combinations of covalent, ionic and van der Waals bonds that exist within the structures. While the coordination arrangement of metals is not altered by slip, new neighbours normally result from slip in ceramics. In the latter case, slip has to be achieved by breaking the strong ionic and covalent bonds, which is more difficult than breaking the metallic bonds. The difficulty of slip in ceramic materials makes them strong but brittle. The true compressive strength and the theoretical tensile strength are very high, provided no porosity or structural defect is present. In practice, however, structural irregularities such as microcracks, grain boundaries and microporosity are difficult, if not impossible, to eliminate. While in ductile materials the stress concentrations around these structural defects may be relieved by plastic flow of the material, this is not possible in brittle materials and fracture occurs at stresses much lower than the theoretically possible strength. Once started, the fracture propagates readily under tension, because the stress concentration is intensified as the crack proceeds. Under compression, a crack will not be self-propagating as loads can be transferred across it; therefore brittle materials are usually much stronger in compression than in tension.

The difference in tensile and compressive strengths of ceramic materials is an important design parameter. The ratios between tensile strength, modulus of rupture and compressive strength can be roughly taken as 1:2:10. If in substituting a ceramic material for a ductile material the same configuration is used, the ceramic component is likely to fail in service or even during assembly. If the potentialities of ceramic materials are to be realised in practice, special designs should be adopted to make use of their advantages and avoid their limitations. The main limitations on the use of ceramic materials in engineering applications are their brittleness, variability of properties, low thermal and mechanical shock resistance, and lack of reliability under tensile stresses.

3.4 PHYSICAL AND CHEMICAL PROPERTIES OF CERAMIC MATERIALS

Ceramic materials generally have higher melting points than those of commonly used metals, and their thermal conductivities fall between those of metals and those of polymers. The thermal conductivity of refractories depends on their composition, crystal structure and texture. Simple crystalline structures, as in silicon carbide, usually have higher thermal conductivity. The variation of thermal conductivity with temperature usually depends on whether the material is crystalline or non-crystalline. For example, fireclay bricks show an increase of thermal conductivity with rising temperature, whereas the more crystalline forsterite and some high aluminas show a decrease with rising temperature.

Thermal shock resistance is a function of thermal conductivity and expansion coefficient in brittle materials. Ceramics generally have lower thermal expansion coefficients than do metals and polymers, but there is a wide range of variation between different types and grades. Ceramics with a lower thermal expansion coefficient and higher thermal conductivity, usually exhibit better thermal shock resistance.

Ceramics are generally non-conducting to electricity and are considered to be dielectric. Most porcelains, aluminas, quartz, mica and glass have volume electrical resistivity values greater than 10^{15} ohm·cm and dielectric constants of up to 12. The electrical resistivity of many ceramics decreases with the introduction of impurities in their structures.

Almost all ceramic materials are resistant to chemical attack, except by hydrofluoric acid and, to some extent, by hot caustic solutions. They are not affected by organic solvents. The specific gravities of most ceramic materials range from 2 to 3, which makes them comparable to those of light metals.

3.5 WHITEWARES

Whitewares include the various types of pottery, china, tile and porcelain. The raw materials used in making whitewares are clay, quartz and feldspar. Oxides, colouring additives and fluxes may also be added. The clays used are generally composed of hydrated aluminium silicate, alumina and silica in different ratios. Clays become plastic when mixed with water and act as a bonding material on drying. The finer the particle size, the stronger and more plastic the clay. However, plasticity and shrinkage increase together, and shrinkage is usually limited to 12% for control of dimensions and warping. Shrinkage may be reduced by the addition of non-plastic, non-shrinking materials such as fired clay, ground flint or similar minerals. The second raw material involved in making whitewares is quartz, or sometimes flint. The addition of quartz reduces shrinkage and improves the component strength in the dry and fired condition. The third constituent of whitewares is feldspar which is a class of alkali–alumina–silicate mineral. In the unfired body the feldspar helps in reducing shrinkage, and during firing, it acts as a flux which dissolves the clay and then the quartz. In addition to feldspar, other fluxing ingredients like $CaCO_3$, $BaCO_3$, $Na_2O \cdot Al_2O_3 \cdot 2SiO_2$, ZnO and Na_2CO_3 may also be added.

Besides their household applications in tiles and sanitary ware, whitewares have many industrial applications in electrical insulation. Electrical whitewares, porcelains, are conventionally divided into low-voltage insulators, high voltage insulators and high-frequency insulators. Low-voltage insulators are suited for applications involving up to about 500 volts, and may be used in the unglazed or glazed condition. Cordierite

TABLE 3.1

PROPERTIES OF SELECTED WHITEWARES

Property	*Cordierite*	*Zircon*	*Steatite*
Physical properties			
Melting point (K)	1 762	1 840	1 840
Maximum service temperature (K)	1 290	1 380	1 290
Coefficient of thermal expansion ($\times 10^6$/K)	2·7	3·8	8·0
Specific gravity	1·9	3·6	2·6
Mechanical properties			
Tensile strength (MN/m^2)	28–56	35–77	35–70
Compressive strength (MN/m^2)	350–670	420–700	450–630
Modulus of elasticity in tension (MN/m^2)	49 000	147 000	105 000
Hardness (mohs)	7	8	7·5
Electrical properties			
Volume resistance (ohm·cm)	$> 10^{14}$	$> 10^{14}$	$> 10^{14}$
Dielectric strength (V/mm)	5600–9200	2400–1200	1000
Dielectric constant (1 megacycle)	5·1	9·3	5·8
Power factor, % (1 megacycle)	0·45	0·27	0·17

($2MgO \cdot 2Al_2O_3 \cdot 5SiO_2$) is widely used in low-voltage applications and Table 3.1 lists its relevant properties. High-voltage whitewares are suitable for voltages higher than 500. These grades are vitrified bodies that are usually glazed for weathering resistance. Zircon porcelain is used extensively in high-voltage applications, and consists principally of zircon with some clay and alkaline fluxes. Zircon porcelain has superior properties to normal porcelain, as shown in Table 3.1. High-frequency applications require low power loss, high dielectric strength and high resistivity. Steatite whiteware is usually used for such applications, although oxide ceramics are also used, as will be discussed later in this chapter. Steatite is essentially composed of talc and a little clay. It has excellent electrical properties and a low cost, but its poor thermal shock resistance and narrow firing-temperature-range limit the size and shape of its products. Table 3.1 lists some of the properties of steatite whiteware.

3.6 REFRACTORIES

The most widely used common refractories are the alumina–silica type, with compositions ranging from nearly pure silica to nearly pure alumina, together with some impurities such as iron and magnesium oxides. The structure usually consists of a glassy matrix binding the crystalline constituents. Silica bricks are made of at least 95% SiO_2. Alumina is one of the most important commercial oxides. At moderately high temperatures, its strength is high, it is chemically stable, and it can be used either in oxidising or reducing atmospheres up to about 2200 K for short periods. Pure alumina refractories withstand higher temperatures than those that incorporate other binders. Table 3.2 lists typical properties and applications of some frequently used alumina ceramics.

Beryllia combines good electrical insulation with high thermal conductivity which makes it useful for special electronic applications. Beryllia is stable to about 2000 K in air, hydrogen, carbon monoxide and vacuum. Its resistance to chemical attack at high temperatures makes beryllia suitable as a crucible material for the melting of high-purity metals and alloys. However, it is costly and difficult to work with and its dust particles are toxic. Table 3.3 lists some properties of beryllia.

Magnesia is not as widely used as alumina or beryllia. It is susceptible to thermal shock, and is less stable than alumina when in contact with most metals at temperatures above 2000 K in reducing atmospheres or in vacuum. Table 3.3 lists some properties of magnesia ceramics.

Pure zirconia suffers an inversion in crystal form at about 1270 K, which causes a 7% contraction in volume with accompanying cracking. When stabilised in a cubic form with about 5% CaO, zirconia can be used at elevated temperatures and can be repeatedly heated above 2500 K without reversion. Zirconia is not wetted by most metals, and it is used for crucibles for melting platinum, palladium, ruthenium and rhodium. It is also used for refractory bricks and extrusion dies. Slags, however, react severely with stabilised zirconia. Table 3.3 lists some properties of zirconia refractories.

Thoria has the highest melting point of any oxide, 3600 K, and is stable under most conditions. Thoria crucibles have been used for melting titanium. Its disadvantages are high cost, and sensitivity to thermal shock due to its high coefficient of thermal expansion and low thermal conductivity. Thoria is also radioactive. Table 3.3 lists some properties of thoria.

Carbides, as a family, contain materials with the highest melting points of all engineering materials. Hafnium and tantalum carbides both have melting

TABLE 3.2

PROPERTIES AND APPLICATIONS OF SOME ALUMINA CERAMICS

Nominal composition Al_2O_3(%)	*Specific gravity*	*Hardness (R45N)*	*Tensile strength (MN/m^2)*	*Compressive strength (MN/m^2)*	*Modulus of elasticity (GN/m^2)*	*Dielectric constant (1kHz a.c.)*	*Volume resistivity (ohm·cm)*	*Applications*
85	3·39	73	155	1 930	221	8·2	$> 10^{14}$	All round ceramic for electrical and mechanical applications. Wear resistant brick
90	3·60	79	221	2 482	276	8·8	$> 10^{14}$	Tough, fine grained ceramic for demanding mechanical applications
94	3·62	78	193	2 103	283	8·9	$> 10^{14}$	Good for metallising, and for all but the most critical electrical and mechanical applications
96	3·72	78	193	2 068	303	9·0	$> 10^{14}$	Excellent for special electronic and mechanical applications

(*Contd.*)

TABLE 3.2 (*Contd.*)

Nominal composition Al_2O_3*(%)*	*Specific gravity*	*Hardness (R45N)*	*Tensile strength* (MN/m^2)	*Compressive strength* (MN/m^2)	*Modulus of elasticity* (GN/m^2)	*Dielectric constant (1kHz a.c.)*	*Volume resistivity (ohm·cm)*	*Applications*
99·5	3·89	83	262	2 620	372	9·8	$> 10^{14}$	Extremely low loss. Used in electronic and some mechanical applications. Thin-film substrates
99·9	3·96	90	310	3 792	386	9·9	$> 10^{15}$	Hard, strong, ultra-pure ceramic for use in severe mechanical applications and hostile environments. Metal cutting tools, dies, labware

TABLE 3.3

SOME PROPERTIES OF OXIDE CERAMICS

Property	*Beryllia 98% BeO*	*Beryllia 99·5% BeO*	*Magnesia*	*Zirconia*	*Thoria*
Specific gravity	2·90	2·88	3·60	5·5–6·10	9·69–10·03
Maximum service temperature (K)	2 690	2 690	2 690	2 790	3 000
Hardness (Knoop)	1 300	1 300	700	1 100	700
Tensile strength (MN/m^2)	85	98	140	147	52
Compressive strength (MN/m^2)	1 980	2 100	840	2 100	1 400
Modulus of elasticity (GN/m^2)	230	245	280	210	140

points of 4217 K. However, carbides cannot be used unprotected at high temperatures because of their poor oxidation resistance. The only exception is silicon carbide which can be used at temperatures up to about 1950 K. Silicon carbide combines high thermal conductivity, low thermal expansion and low thermal shock. Boron carbide is best known for its extreme hardness and its abrasion resistance. Table 3.4 lists the properties and applications of some commonly used carbides.

Most nitrides are relatively brittle and have poor resistance to oxidation. The most widely used refractory nitrides are boron nitride (BN) and silicon nitride (Si_3N_4). Cubic boron nitride is best known as the synthetic diamond material, Borozon, and is stable in air up to 2000 K and has a hardness equal to that of diamond. It also has a thermal conductivity 5 times that of copper at room temperature, which makes it useful for very small heat-sink devices. Boron nitride is anisotropic, and its thermal expansion parallel to the direction of pressing is 10 times that in the perpendicular direction. Silicon nitride has outstanding resistance to wetting or reaction with molten non-ferrous metals, and is used for crucibles and boats for melting and refining semiconductor materials like germanium. It has excellent thermal shock resistance, which makes it useful for thin, thermocouple protection tubes and radiant heat shields. Its excellent erosion resistance at high temperatures makes it useful for rocket nozzle inserts.

TABLE 3.4

PROPERTIES AND APPLICATIONS OF SOME CARBIDES

Material	*Specific gravity*	*Coefficient of thermal expansion* ($\times 10^6/K$)	*Compressive strength* (MN/m^2)	*Modulus of rupture* (MN/m^2)	*Hardness (mohs)*	*Applications*
SiC, silicate bond	2·57	4·68	105	21	—	Excellent refractory for many purposes. Service temperature up to 1990 K in oxidising atmospheres. Used for high temperature furnaces and combustion tubes
SiC, silicon nitride bond	2·62	4·68	140	44	—	More refractory than silicate bonded, does not soften up to 2220 K. Used for more demanding applications
SiC, self-bonded	3·10	3·78	1 400	175	9·2	High thermal conductivity, high thermal shock resistance, outstanding abrasion resistance and chemical inertness. Used for high temperature furnaces, retorts, combustion tubes, rocket nozzles and linings for ball mills
Boron carbide	2·51	4·5	2 900	308	9·3	Excellent neutron absorber. Used in atomic power reactors
Tungsten carbide	16	5·95	—	—	9–10	Cutting tips, tools, and other applications requiring wear and abrasion resistance

There are some ceramic materials with unusual properties that have specific use in the electronics industry. Ferrites are mixed metal-oxide ceramics and combine high electrical resistivity and strong magnetic properties. In contrast to metallic magnetic materials, ferrites have high volume-resistivity and high permeability. Barium and lead ferrites are the best known, and have exceptionally high coercive force. They are widely used in permanent magnet motors in automobiles, small appliances and portable electric tools.

Ferroelectric ceramics can convert electrical signals into mechanical energy, such as sound. They can also change sound, pressure or motion into electrical signals. Barium titanate is the most common ferroelectric ceramic and is used in phonograph pick-up crystals and for transducers.

3.7 GLASS

The word ‘glass’ describes any inorganic product of fusion that has been cooled to a rigid condition without crystallising. Glasses include a large family of materials of widely differing composition and physical properties.

Vitreous silica has a high softening temperature and low thermal expansion coefficient, and is thus suitable for use at high temperatures. It also has very good chemical durability and electrical resistance. However, the high temperatures required for melting silica and the restrictions in forming it increase the manufacturing costs, which limits its use in practice.

The addition of alkali and stabilisers like CaO to silica produces glasses with viscosities considerably below that of silica, and these can be melted at temperatures of about 1570 K. Soda–lime–silica glasses form the most important group in terms of tonnage melted and variety of application. They are used for making bottles, jars, window glass, plate or float glass and electric bulbs.

Pyrex glasses are made by replacing most of the soda in soda–lime glass by boric oxide. This reduces the thermal expansion coefficient which improves thermal shock resistance. Pyrex glasses have excellent resistance to corrosion by acids and can be used at moderate temperatures. They are used typicably for laboratory glassware, industrial glass piping and domestic cooking ware.

Lead glasses are composed basically of silica, potash and lead oxide. They have a high refractive index and high electrical resistivity. Glasses of this type are used for neon sign tubes and other applications requiring good electrical insulation. With higher lead content these glasses can be used for electric capacitors and for absorption of X-rays, gamma rays and other forms of higher radiation. Lead glasses are also used for optical purposes where they are commonly called 'flint glasses'.

Typical compositions and properties of some commercial glasses are given in Table 3.5. Many of the mechanical properties of glass are almost independent of chemical composition. The strength of glass, for example, varies considerably according to the method of manufacture, subsequent treatment and, particularly, surface condition. In theory the strength of conventional soda–lime glass should be between 10 000 and 30 000 MN/m^2, whereas in practice ultimate strengths of about 50 MN/m^2 are used for design purposes. The difference between theory and practice is mainly due to flaws which range from sub-microscopic defects to visible scratches.

Improvements in the strength of glass may be obtained by physical or chemical treatments. Physical treatments involve heating the glass to just below the softening point and then quenching it in air. This treatment introduces compressive stresses in the surface layers, which toughens the glass, and it is used for automobile windscreens. Compressive stresses in the surface layers can also be introduced by chemical treatments. These involve coating a high expansion glass with a low expansion glass, by ion exchange or by surface crystallisation. Chemical toughening methods are more expensive than physical methods and so far have been restricted to such components as aircraft windscreens.

Glass ceramics, pyrocerams, are a family of fine-grained, crystalline materials made from special glass compositions by controlled crystallisation. Glass ceramics are non-porous and although not ductile, they have much greater impact strength than commercial glasses and ceramics. Their thermal expansion coefficient varies from negative to small positive values, depending on the composition, and this leads to excellent thermal shock resistance and good dimensional stability. Glass ceramics are useful up to about 1380 K with excellent corrosion and oxidation resistance. They are electrical insulators and are suitable for high-temperature, high-frequency applications.

3.8 PROCESSING OF CERAMIC MATERIALS

The raw materials used for making ceramic components are usually in the form of particles or powder. After mixing and blending the appropriate ingredients, processing is carried out either in a dry, semi-liquid or liquid state, and in either cold or hot condition.

TABLE 3.5

TYPICAL COMPOSITIONS AND PROPERTIES OF COMMERCIAL GLASSES

	Fused silica	*96% silica (Vycor)*	*Soda–lime–silica*	*Boro-silicate (Pyrex)*	*Alumino-silicate*	*Lead-alkali*
Chemical composition (%)						
SiO_2	99·5 +	96	72·6	80·2	57	56·5
Al_2O_3			1·6	2·6	20·6	1·4
K_2O			0·9	0·3		8·25
Na_2O			13·1	4·5	1·0	4·25
CaO			3·7	0·1	5·4	
MgO			8·0		12·0	
B_2O_3		4		12·3	4	
PbO						29·6
Fe_2O_3			0·1			
Properties						
Young's modulus (GN/m^2)	73·5	67·2	70	66·5	89·6	63
Poisson's ratio	0·17	0·18	0·24	0·20	0·26	—
Specific gravity	2·20	2·18	2·46	2·23	2·53	3·04
Linear expansion coefficient ($\times 10^7$/K)	5·6	8	92	32·5	42	90
Maximum service temperature (K)	1 470	1 370	733	763	910	653
Volume resistivity (ohm·cm) at 523 K	$1{\cdot}6 \times 10^{12}$	5×10^9	$2{\cdot}5 \times 10^6$	$1{\cdot}3 \times 10^8$	$3{\cdot}2 \times 10^{13}$	8×10^8
Dielectric constant	3·8	3·8	7·2	4·6	6·3	6·7
Refractive index	1·458	—	1·510	1·474	1·634	1·583

Slip casting consists of suspending powdered raw materials in liquid, to form a slurry or slip that is poured into porous moulds, usually made of gypsum. The mould absorbs the liquid, leaving a layer of solid material on the mould surface. For hollow components, the excess slip is removed after the desired shell thickness has formed. The process is especially economical for short production runs. Products made by slip casting are liable to large shrinkage on firing because the green density obtained by this method is only about 70% of the theoretical density. Large and complex parts can be easily produced by slip casting. The main disadvantages of this process are poor dimensional accuracy and slow rates of production.

Jiggering is used for circular or oval shapes like dinnerware and procelain electric insulators. In jiggering, the plastic ceramic body is rotated while a profiling tool forms the surface by cutting away excess material. This process can be adapted for automatic machines and high rates of production.

Extrusion consists of forcing a plastic ceramic body through a forming die and then cutting the product to length. Most clay ceramic products such as brick and rain-tile are made with auger extruders. The advantages of extrusion are low tooling costs and high production rates; but it is difficult to produce parts with thin walls, and the parts must be symmetrical.

Pressing can be done with dry, plastic or wet raw materials. In dry pressing, ceramic mixtures with liquid levels up to 5% by weight are pressed into shape under high pressure in a metal die. This method is widely used for manufacturing non-clay refractories, electrical insulators, and electronic ceramic parts. Small parts can be produced to close tolerances by this method. Semi-dry pressing and wet pressing, in which water content is from 5 to 15% and from 15 to 20% respectively, use lower pressures and less expensive dies. In isostatic pressing, dry ceramic powder with small amounts of binder are pressed and then sintered to densities up to about 95% of theoretical densities. The process is widely used for high grade oxide ceramic components of complex shape. Hot pressing combines the pressing and firing operations, and is used with isostatic or uniaxial pressing techniques. Components produced by hot pressing have high densities and strength.

Moulding processes in ceramics are similar to the injection moulding of plastics. The ceramic is mixed with a thermoplastic resin and heated to provide the fluidity needed for the mixture of flow into the mould cavity. The resin is later burned off before firing. Sections as thin as 0·5 mm and as thick as 8 mm can be injection moulded, and complex shapes with good dimensional accuracy can be rapidly produced. The main limitation is high tooling costs.

After forming the plastic ceramic mass into the required shape by one of the above processes, it is dried to remove the water. If drying is carried out too quickly, cracking or warping may occur. After drying, the part is fired or sintered into a permanent product. In whiteware products, firing takes place at temperatures between 1200 and 1700 K. During firing, vitrification takes place and the resulting liquid, glassy phase fills the pores, thus forming the matrix. This holds the unmelted particles together. In high grade refractories, the sintering process produces a crystalline bond instead of the glassy phase bond resulting from vitrification.

Glass can be cast, rolled, drawn or pressed like metallic materials and, in addition, it can be blown. Casting is accomplished by pouring the liquid glass into a mould and cooling slowly. Rolling is widely used for window glass and plate glass. It is accomplished by passing glass of the right viscosity between rollers. The resulting sheet is then passed through the annealing furnace. Plate glass can also be produced by floating the molten glass on a bath of molten tin. This method results in mirror smooth surfaces that need no further grinding or polishing. Drawing of glass tubing is accomplished in a way similar to rolling, in which glass of the proper viscosity flows directly around a ceramic tube or mandrel pulled by asbestos covered rollers. The press and blow method is used to make containers. This process involves feeding a gob into a mould, pressing, and then blowing the glass to take the final shape. The products are than annealed to remove internal stresses caused during cooling.

Glass processing often involves removing material from the glass surface. This can be done by cutting, using carborundum or diamond abrasive wheels. Another method is to cover the surface with an etchant resistant wax, leaving the areas to be marked uncovered. The component is then immersed in a solution containing hydrofluoric acid. Coatings are also frequently applied to glass surfaces during or after processing. A thin film of an inorganic oxide, usually tin or titanium, can be deposited on glass to improve lubricity, thereby improving the scuffing resistance of the surface. The presence of coatings of this type is not apparent to the unaided eye, although surface reflection is increased, and iridescent effects are possible if the film is applied too thickly.

3.9 DESIGN CONSIDERATIONS

Ceramic parts shrink during the firing process, and this shrinkage should be taken into consideration in the design. If closer tolerances are to be achieved after firing, wet grinding at high speeds is usually performed. In cylindrical grinding, accuracies of $\pm$ 0·1 mm can be held, and parallel grinding gives an accuracy of $\pm$ 0·0003 mm per mm.

Refractory ceramics are brittle, and points of stress concentration, such as holes, notches and sharp corners, should be avoided. Tolerances should be generous so as to avoid the need for machining, which is frequently difficult. On coated ceramics, the thickness will thin out on sharp corners, so they should be avoided if uniform thickness is required.

Ceramics are weaker in tension than in compression and designs should avoid stressing in tension. Components formed by moulding should have adequate draft and liberal radii, so that the moulded part may readily be removed from the mould.

BIBLIOGRAPHY

1. H.R. CLAUSER, *Industrial and Engineering Materials*, McGraw-Hill (New York, Maidenhead England), 1975.
2. Z.D. JASTRZEBSKI, *The Nature and Properties of Engineering Materials*, 2nd Edn, John Wiley and Sons (New York), 1976.
3. W.D. KINGERY, H.K. BOWEN and D.R. UHLMANN, *Introduction to Ceramics*, 2nd Edn, John Wiley and Sons (New York), 1976.
4. B.W. NIEBEL and B. DRAPER, *Product Design and Process Engineering*, McGraw-Hill (New York, Maidenhead England), 1975.
5. F.H. NORTON, *Elements of Ceramics*, Addison-Wesley (Reading Mass., London, Amsterdam), 1957.
6. D. RIDER, *Glasses*, Borax Consolidated Ltd (London), 1965.

4

Materials of Construction

4.1 INTRODUCTION

Materials used by the engineer for constructional purposes cover a wide variety of metallic and non-metallic materials. The most widely used metallic material of construction is plain carbon steel, because it compares favourably with other materials on the basis of cost per weight, cost per volume and cost per unit of strength. The versatility of steel, and its availability in a wide variety of standard forms with reproducible properties, are also among the attractive characteristics that make steel so popular as a construction material. The various types and properties of steel were discussed in Chapter 1. Non-ferrous metallic materials are also used for constructional purposes when special requirements justify their relatively higher costs.

Non-metallic construction materials include building stone, clay products, timber, cementing materials, concrete and soil. These materials have complex structures that are not easily characterised but nevertheless affect their behaviour in service. The abundance of these non-metallic materials, their relatively low cost, and their ease of manufacture are the main reasons for their wide use in construction. The characteristics and uses of the non-metallic materials of construction will be discussed in this chapter.

4.2 BUILDING STONE

Building stone has been used in construction since the first Egyptian dynasty and is still being used in modern architecture. The conditions which govern the selection of stone for structural purposes are cost, fashion, ornamental value and durability. In accordance with geological origin rocks may be classified as: igneous, formed by relatively rapid cooling of molten material from inside the earth as in the case of granite; sedimentary, consolidated from particles of decayed rocks which have been deposited from streams of water, as in limestone and sandstone; and metamorphic, either igneous or sedimentary rocks which have undergone structural change due to pressure or heat, as in marble, quartzite and slate.

Generally, the durability and weathering of building stone depends on atmospheric temperature, humidity and pollution. The life of a rock might have been many thousands of years in Ancient Egypt or Greece when it would not last as many scores of years in an industrial city. If they are too high, the thermal expansion and porosity of stone can adversely affect its durability and mechanical properties. Table 4.1 gives some physical and mechanical properties of some commonly used building stones. Stone is largely non-homogeneous, and the properties of two samples may vary considerably. The values given in Table 4.1 are only meant to illustrate the differences between the different kinds.

The non-homogeneous nature of stone and its low impact strength require that a high factor of safety be used in design. Factor of safety values of 20 are commonly used. Besides the effect of the environment, the durability of stone is dependent upon its own physical properties and chemical composition. The presence of discontinuities such as bedding planes and cracks tends to weaken the stone and provides openings for disintegration. The effect of chemical composition on durability is illustrated by the fact that limestone, which is basically calcium carbonate, can be decomposed by

TABLE 4.1

PHYSICAL AND MECHANICAL PROPERTIES OF BUILDING STONES

Kind of stone	*Specific gravity*	*Coefficient of thermal expansion* ($\times 10^{-7}/K$)	*Porosity* (%)	*Absorption* (%)	*Shore hardness number*	*Crushing strength* (MN/m^2)
Granite	2·66	80	1·5	0·6	90	160
Basalt–Gabbro	2·80	56	2·6	1·1	80	162
Sandstone	2·60	97	10·4	4·8	57	110
Limestone–dolomite	2·70	76	2·1	0·9	54	114
Quartzite	2·65	108	0·5	0·2	90	236
Slate	2·80	80	1·4	0·5	56	129
Marble	2·68	81	0·7	0·3	48	140

acids, while granite, which is basically silicate, is more resistant to chemical action. The harder and stronger building stones are generally more durable but more difficult and expensive to quarry and dress to shape.

Granites are hard crystalline rocks which consist of orthoclase, feldspar and quartz plus small amounts of mica. Their chemical composition is usually within the following percentage limits: silica, 65 to 75; alumina, 12 to 18; potash, 3 to 6; soda, 2 to 5; lime, magnesia and iron oxides, less than 2 each. The size and colour of the crystals influences the use of granite. Fine grained granite is suitable for carving and polishing, medium grained qualities of granite are suitable for building construction, while coarse grained qualities are only fit for crushing.

Basalt is a black, imperfectly crystallised igneous rock which consists mainly of augite and plagioclase feldspar. Owing to its sombre colour, its toughness, and difficulties in quarrying and cutting it, little use has been made of basalt for building construction. When crushed it provides good road making material and aggregate for concrete.

Limestone is the trade name for all stratified rocks, which consist principally of calcite or a combination of calcium and magnesium carbonates. In the majority of limestones the content of lime plus magnesium carbonate is more than 75%. Impurities consisting of clay, flint, sand, iron carbonate and oxide, gypsum or alkali carbonate are usually found. Limestone is one of the weakest and softest of building stones, but it is also cheap.

Sandstone consists mostly of quartz grains cemented together by silica, clay, iron oxide or lime carbonate. The colour, hardness, strength and durability of sandstone depend on the binding material. Silica is the most enduring binder and lime carbonate is the poorest. Quartzite is a hard, metamorphic sandstone containing a siliceous cement.

Marble is a metamorphic rock of a limestone origin. Most of the more desirable marbles are finely crystalline rocks that are capable of taking a polish. Slate is another metamorphic rock, of a shale origin, although a few slates have an igneous rock origin. Most slates contain from 55 to 70% silica, 9 to 25% alumina, with small percentages of iron oxide, lime, magnesia and the alkalies. A large majority of the slate produced is used for roofing and electrical purposes.

4.3 CLAY PRODUCTS

Clay products form one of the most important classes of structural materials. Structural clay products may be classified as follows: brick for building, paving or firebrick; tile for roofing, floors, walls and load-bearing; and pipe for sewers, drains and conduits. The minerals commonly found in clays are mostly kaolinite ($2SiO_2 \cdot Al_2O_3 \cdot 2H_2O$) and other hydrated silicates of alumina. Silica forms 40 to 80% of the raw materials used in making structural clay products other than firebrick. In firebrick the silica content may be as high as 98%. The alumina content ranges from 10 to 40% except in silica brick. Iron oxide, which usually constitutes less than 7%, determines the colour of the clay and of the fired product. Lime normally constitutes less than 10% of the clay. Alkalies form less than 10% but are of great value as fluxes.

Plasticity, tensile strength, texture, shrinkage, porosity, fusibility and colour after firing are the important physical properties that determine the value of a clay. The methods of processing clay products are similar to those described for ceramic materials in Chapter 3.

The essential requirements for building brick are sufficient strength in crushing and bending, durability, a proper suction rate and a pleasing appearance if it is to be used as a facing brick. The requirements for tiles are generally higher than those for brick, and better clays and more accurate processing methods are employed. Wall tiles are usually finished by glazing. Clay pipes used for sewers should successfully withstand the action of acids and should be impervious. They should be reasonably straight, smooth and free from cracks and blisters.

4.4 CEMENTING MATERIALS

The cementing materials most widely used in construction may be classified as follows: hydraulic cements, which include portland cement and high alumina cement; limes, which include quick lime and hydrated lime; and gypsum plaster. Portland cement comprises about 97% of the total production of all hydraulic cements and is available in five grades according to ASTM standard C15. Type I is used for general construction (normal portland cement). Type II is used in concrete that is exposed to moderate sulphate action. Type III is used when a high early strength is required. Type IV is used when a cement with low heat of hydration is required. Type V is used when high sulphate resistance is required. The three fundamental constituents of hydraulic cements are lime, silica and alumina. Most cements also contain iron oxide, magnesia, sulphur trioxide, alkalies and carbon dioxide. The minimum tensile and compressive strengths of 1:3 cement mortars are shown in Table 4.2 for different types of cement.

TABLE 4.2

MECHANICAL PROPERTIES OF 1:3 MORTAR MADE FROM DIFFERENT TYPES OF PORTLAND CEMENT

Property	*Type I*	*Type II*	*Type III*	*Type IV*	*Type V*
Tensile strength (MN/m^2)					
1 day in moist air	—	—	1·9	—	—
1 day in moist air, 27 days in water	2·45	2·28	—	2·1	2·1
Compressive strength (MN/m^2)					
1 day in moist air	—	—	8·75	—	—
1 day in moist air, 27 days in water	21	21	—	14	15·4

Gypsum is a combination of sulphate of lime and water of crystallisation. As a structural material gypsum plaster is light with good fire resistance and of low cost. Gypsum plasters are used for wall finishing and for fireproofing around columns. The compressive strength of gypsum varies from about 0·5 to 21 MN/m^2 depending upon the amount of water used in mixing the gypsum paste, the completeness of drying and the temperature used in calcining the gypsum rock. The highest strength is obtained with the least mixing water. Gypsum–cement compounds have higher strength than gypsum.

The basis of the cementing action of lime mortar is the carbonation of the calcium hydroxide formed by the reaction of water with calcium oxide (quick lime). Lime mortars need air to harden and are mainly used for binding brick, concrete and stone masonry, and for plastering interior walls.

4.5 CONCRETE

Concrete is a mixture of crushed stone, gravel or similar inert material, with mortar. The cement is the most expensive ingredient in concrete. Hence, from the standpoint of economy, it is desirable to use as little cement as possible. By properly proportioning the cement, water and aggregates it is often possible to effect considerable savings and still produce a concrete which will fulfil all the requirements for durability, strength, abrasion resistance and watertightness. For well-graded aggregates, the proportions by dry, loose volume shown in Table 4.3, will generally meet the

TABLE 4.3

STRENGTH OF SOME CONCRETE MIXES AFTER 28 DAYS

Structure	*Proportions by volume*			*Strength (MN/m^2)*			
	Cement	*Fine aggregate*	*Coarse aggregate*	*Compressive*	*Modulus of rupture*	*Tensile*	*Modulus of elasticity (compression)*
Columns, structures carrying high stresses	1	1·5	3	40	5	3·2	25 000
Floor slabs, beams, structures carrying ordinary stresses	1	2	3·5	18	3·6	1·7	22 000
Filling and massive work	1	3	5	10	2·3	1·1	19 000

requirements of the types of structures indicated. Water is usually added in proportion to the cement and should be enough to make the concrete plastic and workable. Concrete sets and hardens as a result of the chemical reactions which take place between the compounds of the cement and the mixing water. The strength is affected by the curing conditions, and wet concrete surfaces at moderate temperatures yield the highest strength. Strength usually continues to increase with age in concrete, and strengths at 1 year are about 1·3 to 1·8 times those at 28 days, depending on the mix. Under conditions of use where concrete receives no additional moisture after placing, little gain in strength can be expected after the uncombined mixing water has evaporated. Approximate strength values are given in Table 4.3 for 28-day-old concrete.

Like all brittle materials, concrete is much stronger in compression than in tension. The tensile strength of concrete is usually neglected except for massive work like heavy foundations and walls, piers, arches or massive dams where concrete is used plain, without reinforcement. Where tensile stresses exist in a concrete structural member, the strength to resist them is furnished by steel bars embedded in the concrete. Such construction is known as reinforced concrete. Pre-stressed concrete is a form of reinforced concrete in which high strength steel reinforcement is put under an initial stress to induce pre-compression in the concrete to counteract the tensile stresses produced by external loads. Some of the principal examples of the uses of reinforced concrete are for beams, columns, footings, roof slabs, bridges, tanks and retaining walls.

The properties which are generally most desired in concrete are durability, watertightness, strength and resistance to abrasion or wear. Concrete exposed to sulphate and sea waters must be able to resist the action of the chemicals in these waters. Concrete shrinks and expands with loss or absorption of moisture. These changes in dimensions are independent of any changes which may be caused by temperature or stress. The major proportion of the volume changes with moisture are due to the hardened cement paste. Consequently, the greater the quantity of paste in the mix, the greater will be the expected volume change. The volume changes taking place after the concrete has hardened should be taken care of in the design of the structure by the proper spacing of joints and the arrangement of reinforcing bars. There is little danger of the concrete spalling due to contraction and expansion caused by temperature changes, because the coefficient of thermal expansion of concrete is nearly the same as that of steel.

4.6 SOILS

From an engineering point of view, soils are loose or semi-indurated materials of the Earth's mantle. They are vital to the structural engineer because he must use them for foundations, or in road beds, retaining walls, etc. Soils are most readily characterised on the basis of sand, silt and clay content.

The colloidal clays may be platelike, and possess significant surface charges around their edges as a result of their structures. Therefore, the presence of a clay markedly affects the properties of the soils, making them plastic when wet, reasonably strong when dry and impermeable to liquids.

4.7 TIMBER

Timber has been used as a structural material since the dawn of history. Today it is used because it is cheaper, lighter and more easily shaped than most other construction materials. On the other hand, it is subject to decay and attack by certain insects, and it is also inflammable.

There are numerous kinds of commercial woods, which can be broadly classified into soft woods and hard woods. Soft woods usually come from trees with needlelike leaves, like pine, cedar, fir, spruce, hemlock and cypress. Hard woods usually come from trees with broad leaves like oak, maple, walnut, hickory, ash and poplar.

More than 50% of the structure of wood is made up of cellulose fibres, cemented together by about 10 to 35% lignin. Although the different species of wood have about the same percentage of cellulose and lignin, their mechanical properties differ considerably because of differences in structure. The specific gravity of wood is related to its mechanical strength, and heavier woods are also stronger. The moisture content of wood also was a marked effect on its strength, and increase in moisture causes a decrease in strength.

The structure of wood is anisotropic and this is reflected in its physical and mechanical properties. Thermal expansion is about 40% greater in the tangential direction than in the radial direction, and 6 to 8 times greater in the radial direction than in the logitudinal direction. The tensile Young's modulus in the longitudinal direction is about 20 times that in the tangential or the radial directions. Shrinkage is also anisotropic and longitudinal changes are much smaller than tangential changes; while radial shrinkage is intermediate. This anisotropy in shrinkage leads to distortion and warping on drying. Average properties of some common types of timber are given in Table 4.4.

TABLE 4.4

AVERAGE PROPERTIES OF SELECTED COMMERCIAL TIMBERS

Timber	*Specific gravity*	*Bending*			*Compression parallel to grain*		*Compression perpendicular to grain*	*Ultimate shear parallel to grain (MN/m^2)*
		Proportional limit (MN/m^2)	*Modulus of rupture (MN/m^2)*	*Modulus of elasticity (MN/m^2)*	*Proportional limit (MN/m^2)*	*Ultimate strength (MN/m^2)*	*Proportional limit (MN/m^2)*	
Ash (white)	0·58	62·3	102·2	11 760	39·0	51·0	10·6	13·4
Cedar (red)	0·47	26·6	61·6	6 160	—	42·0	8·0	—
Cypress	0·46	50·4	74·2	10 080	33·2	44·5	6·3	7·0
Elm	0·50	53·2	82·6	9 380	28·2	38·6	6·0	10·6
Hemlock	0·40	42·7	62·3	8 400	28·1	37·9	5·6	7·4
Hickory	0·73	76·3	137·9	15 260	—	62·8	16·2	15·0
Maple	0·54	60·9	93·8	11 480	32·6	45·8	8·7	13·0
Oak (red)	0·63	58·8	101·0	12 670	32·3	48·4	8·8	12·8
Oak (white)	0·67	55·3	97·3	11 340	30·5	49·3	9·9	13·2
Pine (white)	0·36	42·0	61·6	8 960	25·8	33·9	3·9	6·0
Pine (yellow)	0·58	65·1	102·9	13 930	43·1	59·1	8·3	10·5
Redwood	0·40	48·3	70·0	9 380	31·9	43·1	6·0	6·6
Spruce	0·40	46·9	71·4	10 990	33·5	39·3	5·0	8·1

Except where designs load timber in the longitudinal direction, anisotropy is undesirable and techniques have been developed to overcome it. Examples include plywood in which the layers are placed at right angles and impregnated wood in which the pores are filled with a polymer like phenol formaldehyde to provide better bonding.

Deterioration of the properties of wood is usually caused by various bacteria and fungi. Termites, beetles and ants damage wood by tunnelling through it. Decay of wood can be decreased by the use of preservatives like coal tar or zinc chloride. The fire resistance of wood may be increased by using surface treatments or by impregnation. Special paints and various chemicals are used for these treatments.

BIBLIOGRAPHY

1. Z.D. Jastrzebski, *The Nature and Properties of Engineering Materials*, 2nd Edn, John Wiley and Sons (New York), 1976.
2. H.F. Moore and M.B. Moore, *Textbook of the Materials of Engineering*, McGraw-Hill (New York, Maidenhead England), 1953.
3. H.W. Pollack, *Materials Science and Metallurgy*, Reston Pub. Co., Prentice-Hall (New Jersey), 1973.
4. H.F.W. Taylor, *Progress in Ceramic Science*, Vol. 1, J.E. Burke, Ed., Pergamon (Oxford, New York), 1961.
5. L.H. Van Vlack, *Elements of Materials Science*, 2nd Edn, Addison-Wesley (London), 1964.
6. M.O. Withey and G.W. Washa, *Materials of Construction*, John Wiley and Sons (New York), 1954.
7. J.F. Young, *Materials and Processes*, 2nd Edn, John Wiley and Sons (New York), 1954.

5

Multiphase Materials

5.1 INTRODUCTION

In many engineering products, different types of materials are often combined in a single component in order to optimise its properties and behaviour under service conditions. Each of the contributing materials usually serves one or more specific functions in the finished product. The overall behaviour of multiphase materials is even more dependent on their structures than is the behaviour of single phase materials. In multiphase materials the properties are affected by the size and distribution of the phases in relation to each other, the bond strength between the different phases, the shape and amount of each phase and the properties of each phase.

The different materials in a multiphase material may be present as macroscopically distinct phases, as in the case of coatings, sandwich materials and laminates, or they may be mixed on a microscopic scale, as in the case of dispersion, particulate and fibre composites. By this definition most engineering materials are multiphase; for example, in metallic materials all ferrous alloys and most non-ferrous alloys are multiphase materials.

Most multiphase materials have been developed to improve mechanical properties such as strength, stiffness, creep resistance and toughness, but a few examples have been developed to achieve certain physical or chemical characteristics. Some properties of multiphase materials depend only on the amount of each phase and are insensitive to the microstructural geometry. These microstructure-insensitive properties may be determined by suitable weighted averages of the properties of each of the individual phases. Density and heat capacity are examples of such properties. The variation of microstructure-insensitive properties with the amount of phases can be represented as:

$$P_c = f_1 P_1 + f_2 P_2 + \ldots \quad (5.1)$$

where P_c, P_1 and P_2 are the properties of the multiphase, phase 1 and phase 2 materials respectively, and f_1 and f_2 are the volume fractions of phases 1 and 2 respectively.

Multiphase material properties which are sensitive to the geometry of the phases as well as to their amounts are called 'structure-sensitive' properties. Examples are elastic modulus and thermal and electrical conductivities. The variation of these properties with the volume fraction of constituent phases does not follow a unique relationship, but usually falls between an upper limit and lower limit depending on the arrangement of phases. The upper limit is represented by the rule of mixtures of Eqn. (5.1) while the lower limit is represented by the relationship:

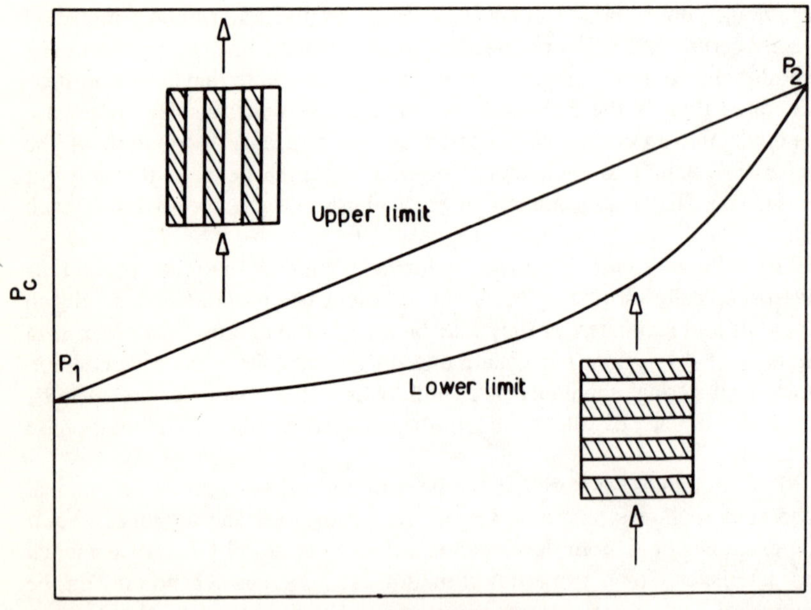

FIG. 5.1. Variation of structure sensitive properties with volume fraction for multiphase materials.

$$P_c = \frac{P_1 P_2}{f_1 P_1 + f_2 P_2} \tag{5.2}$$

The upper limit represents the case where the phases are arranged in parallel, while the lower limit is for the case where the phases are arranged in series in relation to the direction of measurement, as shown in Fig. 5.1.

5.2 DISPERSION COMPOSITES

Dispersion composites are characterised by microstructures consisting of fine particles, of sizes 0·01 to 0·1 μm, dispersed in a single phase matrix. Volume fractions usually range from 1 to 15% for the dispersed phase. This type of composite is usually produced to improve the mechanical strength of the matrix where the fine dispersion impedes the motion of matrix dislocations, as discussed in Chapter 1. The main variables in determining the effectiveness of a dispersion are the mean free path between dispersion (*mfp*), the interdispersion separation (*Dp*), and size (*d*) and the volume fraction (*Vp*). The relationships between these parameters are:

$$mfp = \frac{2d}{3\,Vp}(1 - Vp) \tag{5.3}$$

and

$$Dp = \frac{2d^2}{3\,Vp}(1 - Vp) \tag{5.4}$$

Typical ranges for these parameters in dispersion strengthened composite materials are: *mfp*, 0·3 to 0·01 μm; *Dp*, 0·3 to 0·01 μm; *Vp*, 0·01 to 0·15; *d*, 0·01 to 0·1 μm.

Dispersion-strengthening has an advantage over precipitation-hardening because the hard, dispersed phases function as dislocation obstacles at high temperatures while the precipitates go back into solution. Industrial examples of dispersion-strengthened composites are the aluminium–Al_2O_3 system, known as SAP, and the nickel–3% ThO_2 system, known as TD nickel.

For high temperature stability, the dispersed phase must not coarsen which means that it should have low solubility and a low rate of diffusion in the matrix, low interfacial energy with the matrix, low reactivity with the matrix, a high melting point and high negative heat of formation.

Several techniques have been developed for processing dispersion composites. Surface oxidation is the industrial process used for making

SAP, and it relies on the formation of a thin, adherent refractory oxide film on the aluminium powder. The thickness of the oxide film is of the order of 100 A. The oxidised powders are then shaped by conventional powder metallurgy techniques. Internal oxidation is another industrial process and is based on the preferential oxidation of the solute element in a dilute solid solution. This process results in the formation of a fine oxide dispersion throughout the metal matrix. This requires that the solute oxide has a higher negative heat of formation than the matrix oxide. Copper containing 3·5% by volume of Al_2O_3 in fine dispersion is produced by this method. A third process for manufacturing dispersion composites is the co-precipitation process which is based on the chemical precipitation of a nickel salt around very fine thorium oxide particles. This nickel salt is decomposed to nickel oxide and then reduced in hydrogen to produce a nickel powder with a dispersion of thorium oxide; the resulting material is called TD nickel. Figure 5.2 illustrates the behaviour of some dispersion-strengthened composites in relation to other alloy systems.

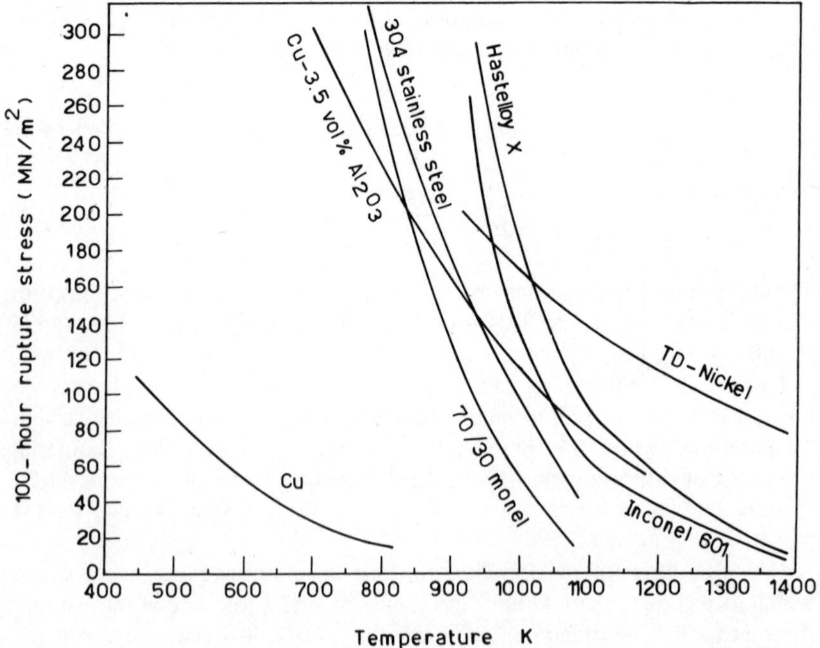

FIG. 5.2. Variation of 100-hour rupture stress with test temperature, for dispersion strengthened Cu–Al_2O_3 and TD–nickel composites, and for some high temperature materials.

5.3 PARTICULATE COMPOSITES

The difference between particulate composites and dispersion composites is in the size of their dispersed phases and of their volume fraction. The particles are more than 1 μm and their volume fraction is usually more than 20%.

Strengthening of particulate-reinforced composites is mostly due to restriction of matrix deformation. The magnitude of this restriction is a function of the interparticle spacing-to-diameter ratio and the ratio of the elastic properties of the particle and matrix. Generally the elastic modulus of particulate composites falls between the upper and lower limits given by Eqns. (5.1) and (5.2) respectively. Positive deviations from Eqn. (5.2) signify matrix constraint.

Beyond the elastic portion of the stress–strain curve of inorganic particulate composites, the softer matrix begins to deform plastically while the harder particles usually remain elastic until fracture. This behaviour imparts strength to the composite but drastically lowers its ductility below that of the matrix phase alone. Typical composites of this type are W–Cu, Mo–Cu, WC–Cu, WC–Co, TiC–Ni, TiC–Ni–Mo and other ceramic–metal combinations, called cermets.

Cermet type composites can be prepared by impregnation, where a green porous compact of the particles is prepared by cold or hot pressing and then impregnated with the liquid matrix material. Conventional powder metallurgy techniques involving either solid state or liquid state sintering are also suitable for the preparation of inorganic particulate composites.

The widest use for cermets is for cutting tools and a detailed study of their characteristics and uses in this field is given in Chapter 14.

The use of particulate fillers in polymeric materials is widespread. Inorganic particulate fillers may be used to raise the elastic modulus, increase the surface hardness, reduce shrinkage and eliminate crazing after moulding, improve fire resistance, improve colour and appearance, modify the thermal and electrical conductivities and raise the viscosity for ease of moulding. Also, the use of inorganic fillers can greatly reduce the cost without necessarily sacrificing other desirable properties.

The addition of spherical carbon black particles, blacks, to rubbers improves their elastic modulus according to the relationship:

$$E_c = E_o(1 + 2{\cdot}5f + 14{\cdot}1f^2) \tag{5.5}$$

where E_c and E_o are the elastic moduli of the filled and unfilled rubber respectively and f is the volume fraction of the filler. With more reinforcing blacks, which are more or less aggregated into chainlike clusters in rubber, higher modulus values than predicted by Eqn. (5.5) are obtained. The effects of different filler materials on the properties of selected rubbers are given in Table 5.1.

TABLE 5.1

THE EFFECT OF DIFFERENT FILLER MATERIALS ON THE PROPERTIES OF RUBBERS

Rubber	*Filler*	*Tensile strength* (MN/m^2)	*Stress at 300% extension* (MN/m^2)	*Abrasion resistance index*
Natural rubber (30 vol. %)	Silica	23·8	5·1	—
	EPC black	29·4	12·0	—
	SAF black	—	12·5	130
	FEF black	—	12·5	100
Styrene butadiene rubber (30 vol. %)	Silica	26·3	7·4	—
	EPC black	28·8	13·5	—
Acrylonitrile butadiene rubber (30 vol. %)	Silica	28·9	9·6	—
	EPC black	16·2	10·2	—

5.4 FIBROUS COMPOSITES

Fibrous composites cover a wide variety of materials in which a matrix is used to bind together the fibres and to protect their surfaces from damage or chemical attack. Furthermore, the matrix separates the individual fibres and prevents brittle cracks from spreading across the composite. The strength of fibrous composites is determined by the strength of the fibres and by the strength and nature of the bond between the fibres and the matrix. The matrix serves as a medium that transfers and distributes the load to the fibres. The bond between the fibre and the matrix must be strong enough to prevent interfacial separation or fibre pull-out under axial loads. Extensive chemical reaction between the fibres and the matrix can damage the fibres and reduce their strength. Such reactions can take place during the processing of the composite or in service, and measures should be taken to prevent them.

In axially-loaded, continuous-fibre composites, the strains in the matrix and in the fibres are the same, and stresses are distributed according to the relationship:

$$\sigma_c = \sigma_m f_m + \sigma_f f_f \tag{5.6}$$

where σ_c, σ_m, and σ_f are the stresses in the composite, the matrix, and fibres respectively, and f_m and f_f are the volume fractions of matrix and fibres respectively. When both the fibres and matrix are deforming elastically, the elastic modulus of the composite (E_{c1}) can be represented by the relationship:

$$E_{c1} = E_m f_m + E_f f_f \tag{5.7}$$

where E_m and E_f are the elastic moduli of the matrix and fibres respectively. Under conditions where the matrix is strained beyond its elastic limit while the fibres are still elastic, the composite modulus, E_{c2}, usually called secondary modulus, is given by the relationship:

$$E_{c2} = E_f f_f + \left(\frac{d\sigma_m}{d\varepsilon_m}\right)\varepsilon_f f_m \tag{5.8}$$

where $\left(\frac{d\sigma_m}{d\varepsilon_m}\right)_{\varepsilon_f}$ is the slope of the stress–strain curve of the matrix at the strain ε_f of the fibres. Continuous fibre composites, containing ductile matrix and brittle fibres, fail when σ_f of Eqn. (5.6) reaches the fracture stress of the fibres, provided f_f is greater than the critical value required for effective fibre strengthening.

When fibrous composites are loaded at 90° to the fibre direction, they fail at stresses as low as the matrix strength or even lower. This anisotropy can be reduced by using the cross-plied arrangement in which the orientation of the fibres is rotated through 90° in the different layers. A more isotropic composite can be obtained if alternate layers are rotated through 45°. However, the composite remains weak in the thickness direction.

The behaviour of composites containing discontinuous fibres cannot be described by relations such as Eqns. (5.6), (5.7) and (5.8) unless their aspect ratio is greater than a critical value that is determined by the relative strength of the fibre and matrix. The higher the difference between the fibre and matrix strengths, the larger the critical aspect ratio. The above equations can be used to anticipate the composite behaviour, provided that the values of σ_f or f_f are adjusted to suitable effective values. The smaller the fibre aspect ratio, the lower the effective fibre strength or volume fraction.

In selecting the matrix and fibre materials to make a composite, the following factors should be considered:

1. The matrix should wet the fibres to reduce the probability of voids at the interface. Fibres are frequently pre-coated with a thin film of a suitable material to improve wettability.
2. Extensive detrimental reactions should not take place between the matrix and the fibre materials. These reactions are usually prevented by fibre coatings. For example, glass fibres are coated with resins, while alumina fibres or whiskers do not require coating when used with the epoxy resin matrix. Metals generally do not wet ceramic whiskers, and metallic coatings are required. In the case of carbon fibres, the bond is mainly mechanical in nature and is determined by the difference in thermal expansion coefficients between the fibres and the matrix, and by interlocking of the matrix with surface irregularities of the fibres.
3. The difference between the thermal expansion coefficients of the fibres and the matrix should not be so large as to result in excessive thermal stresses on cooling or heating the composite. Thermal stresses can adversely affect the properties of the composite, and may cause warping of the finished component. In severe cases, matrix cracking can occur if the local tensile stress exceeds the matrix tensile strength. The differences in thermal expansion coefficients can be used to advantage in hot pressed ceramic matrices with metallic fibres. The fibre material is chosen to have a higher thermal expansion coefficient than the matrix, and on cooling after hot pressing the matrix is placed under compressive stresses in a similar way to pre-stressed concrete.

The materials commonly used in making fibres for reinforcing composites can be divided into metallic fibres or wires, non-metallic fibres and whiskers. Table 5.2 lists some fibre materials and their characteristics.

Glass fibre reinforced plastics account for over three quarters of total fibre composite production. Although many types of plastics can be used as the matrix for glass reinforced plastics, polyester resins are the most widely used. Table 5.3 lists the properties of selected fibre reinforced composites, and Fig. 5.3 shows the variation of specific strength with temperature for different composites.

Filament winding is an established technique for the preparation of continuous, fibre reinforced composites. It involves continuous winding of alternate layers of fibres and matrix material on a mandril of the desired

TABLE 5.2

CHARACTERISTICS OF SOME FIBRE MATERIALS

Fibre Material	*Tensile strength* (GN/m^2)	*Elastic modulus* (GN/m^2)	*Specific gravity*	*Melting temperature* (*K*)	*Range of fibre diameter* (μm)	*Specific strength* (GN/m^2)	*Specific elastic modulus* (GN/m^2)	*Relative cost**
Whiskers								
Graphite	21	686	2·2	3 273	—	9·5	312	—
Al_2O_3	15·4	530	4·0	2 323	3–10	3·85	132·5	7 000
Iron	12·6	196	7·8	1 813	—	1·6	25·1	—
Si_3N_4	14	385	3·1	2 173	—	4·5	124·2	—
SiC	21	700	3·2	2 873	1–3	6·6	219	250–900
Si	7·7	182	2·3	1 723	—	3·3	79·1	—
Non–metallic fibres								
Boron	3·5	420	2·6	2 323	100	1·3	162	1 500
S glass	2·8	84·8	2·49	1 033	10	1·1	34	1
E glass	3·5	73·5	2·54	1 119	10	1·4	28·9	0·5
SiO_2	5·95	73·5	2·19	1 933	35	2·7	33·6	30
Al_2O_3	2·1	175	3·15	2 313	—	0·67	55·6	—
Graphite	2·45	210	1·5	2 923	5	1·6	140	500
SiC	2·1	490	4·0	2 963	76	0·53	122·5	4 000
Metallic fibres								
W	4·1	413	19.4	3 673	13	0·2	21·3	710
Mo	2·2	364	10·2	2 893	25	0·22	35·7	630
René 41	2·0	168	8·26	1 623	25	0·24	20·3	600
Steel	4·2	203	7·74	1 673	13	0·54	26·2	50
Be	1·3	245	1·83	1 453	130	0·71	133·9	10 000

*Relative cost is calculated in relation to the cost of S glass per unit weight which is considered as unity. Relative cost values are based on 1976 US prices.

TABLE 5.3

SOME PROPERTIES OF SELECTED FIBROUS COMPOSITES

Composite system	*Specific gravity*	*Tensile strength (GN/m^2)*	*Elastic modulus (GN/m^2)*
Plastic matrix			
Epoxy + 35% Si_3N_4 whiskers	1·9	0·28	105
Epoxy + 14% Al_2O_3 whiskers	1·64	0·79	42
Epoxy + 70% S glass fibres	2·11	2·1	62·3
Epoxy + 14% S glass fibres	1·38	0·5	—
Epoxy + 70% S glass fabric	—	0·68	22
Polyester + 65% E glass fabric	1·8	0·34	19·6
Metal matrix			
Al + 47% SiO_2 fibres		0·91	
Al + 25% steel fibres		1·2	
Ni + 8% B fibres		2·7	
Ni + 40% W fibres		1·1	
Fe + 36% Al_2O_3 whiskers		1·6	
Cu + 77% W fibres		1·8	
Ta + 29% Ta_2C fibres		1·1	
Co + 30% W fibres		0·7	

shape. After the desired shape is formed, heat and pressure are applied to cure or sinter the matrix material. Other composite manufacturing methods are summarised in Table 5.4.

5.5 LAMINATED COMPOSITES

Laminated composites, or layered materials, consist of two or more different layers bonded together. The layers can differ in material, orientation or form. Clad metals are an example of metal–metal laminates where the outer layers are usually selected to give corrosion resistance or decorative appearance, while the inner layers are usually selected to give strength. Rubber–fabric belts are an example of organic–organic laminates. Other

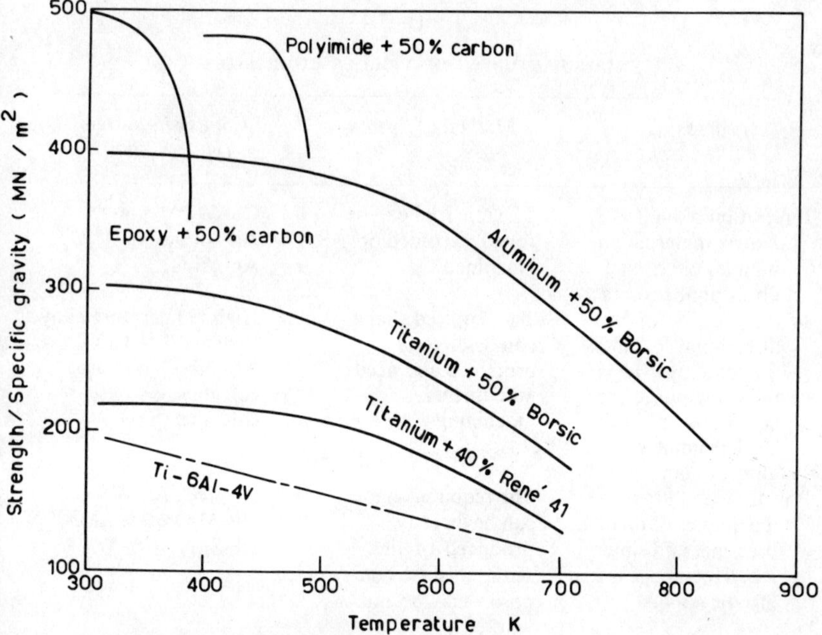

FIG. 5.3. Variation of specific strength with temperature for some fibre-reinforced composites and Ti–6Al–4 V alloy.

possible laminate combinations are metal–organic, organic–inorganic and inorganic–inorganic laminates.

Sandwich materials can be classified as laminated composites. These materials usually consist of a thin facing material and a low density core. Sandwich materials combine high section modulus with low density, e.g. an aluminium-faced, honeycomb sandwich structure beam is about one fifth the weight of a solid aluminium beam of equivalent rigidity. The facing material in a sandwich structure carries the major applied load and therefore determines the stiffness, stability, and strength of the composite. Examples of possible facing materials are aluminium, stainless steel, magnesium, titanium and plastics. The core forms the bulk of a sandwich structure. Therefore, it is usually of light weight but must also be strong enough to withstand normal shear and compressive loads. The core can be in the form of foam or cells. Foam cores are usually made from plastics, especially polystyrene, urethane, cellulose acetate, phenolic, epoxy and silicone. Foamed inorganic materials like glass, ceramics and concrete can also be

TABLE 5.4

MANUFACTURING OF FIBROUS COMPOSITES

Process	*Method of forming*	*Composite system (matrix–fibre)*
Infiltration of molten matrix material into bundles of parallel fibres of fibre mats	The material can be rolled, extruded or machined	Cu–W, Al–glass, Al–Al_2O_3, Ni–Al_2O_3, Ag–Al_2O_3
In situ preparation by directional solidification of alloys. The reforcing phase can be fibrous or lamellar depending on the alloy system	The required shape can be directly produced and needs little further machining	High temperature alloy systems, Ta–Ta_2C, Nb–Nb_2C, Ni base eutectics, Co base eutectics
Powder metallurgy techniques of mixing, pressing, and sintering. Hot pressing can also be applied	The required shape can be directly produced or the semi-finished composite can be hot worked into the finished form	Al–steel, Al–W, Ni–Al_2O_3, Cr–Al_2O_3, Ni–Si_3N_4, Co–W
Electroplating or vapour deposition of matrix on fibres followed by hot pressing	Hot pressed material can be machined to final form	Ni–Al_2O_3, Ni–W
Laminations can be produced by hot pressing of alternate layers of matrix and fibres	Hot pressed sheets can be hot rolled or re-pressed to required shape	Al–steel, plastics laminating systems

used. Cellular cores can have corrugated or honeycomb structure and are usually made from metal foils joined by welding, brazing or adhesive bonding. Other core materials are glass reinforced plastics, ceramics or even paper.

5.6 DESIGN CONSIDERATIONS FOR FIBROUS COMPOSITE MATERIALS

To obtain the maximum strength in one direction in a fibrous composite material, the fibres should be well aligned in that direction. This arrangement of fibres leads to maximum anisotropy in properties, which may be undesirable in some applications. Isotropy can be ensured by cross-lamination of aligned composite layers or by random alignment of the fibres. This latter technique is not practical for cases where large volume fractions of fibres are to be used because of the lower packing efficiency. In applications where a composite material is subjected to tensile loading, the important design criterion is the tensile strength in the loading direction. Under compression loading, failure by buckling becomes an important parameter. For a structural component of a certain configuration, the buckling stress is proportional to the elastic modulus of the component material. In such cases the important design criterion is the composite elastic modulus.

An important factor that should be borne in mind when designing with composite materials is that their high strengths are obtained only as a result of large elastic strains in the fibres. In some structures these strains can cause unacceptable deflections. An example is the case of an aircraft wing where a strain of about 1·7% could cause the wing tips to become vertical. This indicates that composite materials should not simply be substituted for other materials without making the necessary modifications in design and construction.

In spite of their high strength, fibrous composites can exhibit low fracture toughness. This is because the conditions required for high strength, e.g. short critical aspect ratio and strong interfacial bond between fibres and matrix, are in conflict with the conditions required for high fracture toughness, e.g. long critical aspect ratio and weak interfaces parallel to the fibres. In general, a compromise between strength and toughness will be required for a given application.

The method of joining is another important factor that should be considered in the design when using fibrous composite materials. Normal fusion welding techniques cannot be used for fibrous composites without destroying their structure and weakening their mechanical strength. Joining by riveting or bolting is not practical unless the load is evenly distributed near the holes. The stress concentrations usually associated with rivets and bolts can cause local delamination or cracking. Brazing or soldering are not expected to alter the fibrous structure seriously and thus can be used. However, brazed or soldered joints are generally not strong enough for the type of loading that is usually applied to composite materials. In many cases the only satisfactory solution is to fabricate the composite structure in one piece.

5.7 APPLICATIONS OF COMPOSITE MATERIALS

The cost of composite materials is generally higher than the cost of other engineering materials. For example, the cost of Al_2O_3 whisker reinforced composites can be as high as 10 times that of gold. This indicates that composites should be considered as high performance–special purpose materials, and should only be used in cases where their unique properties lead to eventual savings.

The outstanding values for strength/weight and modulus/weight of some composite materials make them prime candidates for aerospace structures. In aerospace applications, lifting a greater payload with a given amount of power can be achieved by reducing the weight of the structure. The same principles also apply to aircraft where higher speeds and altitudes of flight place more severe demands upon the structural materials. Composites with high specific strength at relatively low temperatures are usable for the fuselage and landing gear, and composites that retain their strength and creep resistance at high temperatures are usable for the propulsion system. As an example, the Boeing 747 employs several types of composite materials in several locations as shown in Table 5.5.

TABLE 5.5

COMPOSITE MATERIALS USED IN THE BOEING 747 AIRCRAFT

Location	*Part*	*Composite system*
Wing	Leading edge.	Reinforced glass fibre structure.
	Trailing edge.	Reinforced glass fibre structure.
	Control surfaces (e.g. spoilers, flaps)	Plastic honeycomb core sandwiched in glass fibre with a coating of aluminium
Body	Floor panels	Titanium CP bonded to PVC foam core
		Aluminium bonded to PVC foam core
Vertical fin	Rudder	Glass fibre laminate-honeycomb core
Horizontal stabiliser	Elevators	Glass fibre laminate-honeycomb core
Power plant	Hot thrust reverser	Inconel 625 brazed honeycomb
Strut	Skins	2024 aluminium bonded laminate
Interiors		Glass fibre laminate
Air conditioning	Distribution	Plastic and glass fibre

BIBLIOGRAPHY

1. L. J. Broutman and R.H. Krock, *Modern Composite Materials*, Addison-Wesley (Reading Mass., London, Amsterdam), 1976.
2. D. Cratchley, *Metallurgical Reviews*, **10**(95) (1965), 79.
3. L.W. Davis and S.W. Bradstreet, *Metal and Ceramic Matrix Composites*, Canners Pub. Co., 1970.
4. G.S. Holister and C. Thomas, *Fiber Reinforced Materials,* Elsevier (Amsterdam), 1966.
5. L.Holliday, Ed., *Composite Materials,* Elsevier (Amsterdam), 1966.
6. A. Kelly and G.J. Davies, *Metallurgical Reviews*, **10**(94) (1965), 1.
7. J.B. Moss, *Properties of Engineering Materials,* Butterworths (London), 1971.
8. R.F. Winters, Ed., *Newer Engineering Materials*, Macmillan (New York), 1969.

6

Service Stability

6.1 INTRODUCTION

In many engineering applications, service conditions may produce alterations in material properties which could make them incapable of continuing their function. These alterations can be caused by the chemical environment, heat or radiation. Environments in which most engineering materials serve have a chemically active component that can produce a certain amount of surface damage. These active components can be gaseous, such as oxygen, or liquid, such as chemicals or water. The reaction of the chemical environment with materials results in either a loss of material or a deterioration of surface quality, or both. It can also affect other properties such as wear resistance and fatigue strength. Alterations caused by heat or radiation usually affect the bulk properties of the material, by changing its structure either at the atomic or molecular level, as in radition damage, or at the microstructural level, as in thermal damage.

In selecting a material for a given application, it should be borne in mind that besides performing its mechanical or physical functions, the material should also resist harmful attack of the service environment.

6.2 CORROSION OF METALLIC MATERIALS

Corrosion may be defined as the destructive chemical or electrochemical reaction between a material and its environment. Metals, polymers and ceramics all suffer attack from different atmospheres and solutions. However, the corrosion of metals is electrochemical in nature, while in the other cases a simple solution is usually involved.

Contact between a metal and a liquid solution results in corrosive attack. The degree of attack depends on the metal and the environment, but almost all common metallic materials will corrode under certain conditions. The mechanism of corrosion in aqueous solutions is electrochemical in nature. The corroding metal is the anode in the glavanic cell and the cathode can be another metal, a conducting non-metal or an oxide. Corrosion involving two dissimilar metals, connected together in an electrolyte, is quite common, but the majority of corrosion problems are concerned with one metal only. In these cases, small corrosion cells are set up on the surface of the corroding metal. The electrodes of the cells are usually located at breaks in the surface oxide film or at different phases of the microstructure, e.g. carbides in steel and α–β structures in brasses. Differences of oxygen concentration in the solution at different points on the metal surface, or differences in temperature, stressing or thermal treatment of the metal can also determine the location of the anode and cathode areas of the corrosion cell.

Atmospheric corrosion of metals is probably the most commonly encountered form of corrosion. Any metal exposed to the atmosphere will cover itself with a thin layer of condensed or adsorbed water, even at relative humidities less than 100%, and this layer can act as the electrolyte. The presence of industrial contaminants in the atmosphere increases the corrosion rate. Examples are dust, sulphur dioxide and ammonium sulphate. Sodium chloride is also an impurity, which is present in marine atmospheres, and increases the corrosion rate.

Corrosion of metal surfaces can either take place evenly over the entire surface or can be concentrated at certain points or pits so that the resulting attack is known as pitting. Intergranular attack occurs where the material in the grain boundaries is more susceptible to corrosion than the material of the grains. This type of attack is often strongly dependent on the mechanical and thermal treatment given to the alloy. Unstabilised stainless steels are subject to this type of attack when heated to the temperature range 770–970 K. Intergranular attack also occurs in some nickel alloys when heated in contact with sulphur or its compounds. Stress corrosion cracking arises from the combined effect of static stresses and a corrodent, while corrosion fatigue arises from the combined effect of repeated stresses and a corrodent. The former type of attack can be prevented by interposing a resistant lining or by stress-relief heat treatment, while the latter type is prevented by avoiding notches and other stress raisers. Differential aeration corrosion and crevice corrosion arises from differences in oxygen concentration. In this case metal loss occurs in areas where oxygen concentration is lowest, e.g. at the bottom of the crevice in lap seams of steel structures.

When two different metallic materials are connected in the presence of a corrodent, glavanic corrosion takes place and the severity of the attack depends on the separation of the two materials in the galvanic series (Table 6.1) and on their relative areas. Thus a steel rivet in a copper plate will be more severely corroded than a steel plate containing a copper rivet.

Fretting is a type of corrosion that takes place in tightly fitted parts with slight relative movement, and the corroded surfaces become pitted. Dezincification and graphitisation are forms of selective corrosion where zinc or iron metals are gradually lost from the surfaces of brasses and grey cast irons respectively.

The behaviour of metallic materials under corrosive conditions usually

TABLE 6.1

POSITION OF SOME ENGINEERING METALLIC MATERIALS IN THE GALVANIC SERIES

	Relative position	*Material*
Noble, protected, cathodic	108	Platinum
	107	Gold
	105	Graphite
	100	Silver
	90	Monel (70% Ni–30% Cu)
	89	Titanium
	88	18/8 Cr–Ni austenitic steels (passive)
	85–87	Nickel and Inconel (passive)
	84	Silver solder
	81	Cupro–nickel (70% Cu–30% Ni)
	72–74	Copper
	70	Aluminium bronze
	61–69	Brasses
	62–64	Nickel and Inconel (active)
	53	Tin
	52	Lead
	47	18/8 Cr–Ni austenitic steel (active)
	45	Lead–tin solder
	36	Cast iron
	31–34	Mild and low-alloy steels
	28–29	Aluminium–copper alloys and duralumin
	26	Cadmium
	24	Commercial pure aluminium
	21–23	Aluminium–magnesium alloys
	14	Zinc
Base, corroded, anodic	8	Magnesium and its alloys

differs according to the corrodent, and Table 6.2 can be used for rough estimation of the resistance of some common alloys. The given data are only meant as a guide, and for large production, pilot tests should be carried out to measure the actual performance.

Corrosion at high temperatures can take place in many industrial applications where sulphur is present, either in the material to be heated or in the fuel. Sulphur, in the form of hydrogen sulphide, is found in many operations in oil refineries when the crude oil contains sulphur. Fuels containing sulphur can cause fuel ash corrosion by producing a liquid phase of alkali–metal pyrosulphate compounds, which react with and destroy the normally protective oxide films. Treatment of the fuel with magnesium and aluminium-bearing additives has been used to reduce the problem, as such additives form relatively innocuous compounds that have higher melting temperature.

6.3 CHEMICAL STABILITY OF CERAMIC MATERIALS

As discussed in the previous section, corrosion of metallic materials is strongly related to their electronic structure, as such corrosion is caused by electrochemical processes. Ceramic materials have ionic or covalent bonds that are chemically very stable. The chemical stability of the bonds is reflected in the bulk properties, and the chemical stability of ceramics is one of their outstanding features. Porcelain, for example, resists all acids except hydrofluoric, and it has good stability in even strong caustic alkalies. Stability is retained at elevated temperatures.

Although most refractory ceramic compounds have good resistance to chemical attack, a particular corrosive environment requires careful selection of the ceramic material used. A strong acid environment requires an acidic refractory, whereas alkalies call for a basic refractory material.

Glasses are among the most chemically stable materials. They are exceptionally resistant to attack by water, atmosphere and aqueous solutions of most acids, alkalies and salts. Although the general chemical stability of all glasses is good, their relative performance in various environments may vary considerably between different grades. For example borosilicate and silica glasses show much higher resistance to boiling water and hot dilute acid solution than do soda lime and lead alkali glasses.

6.4 CHEMICAL STABILITY OF POLYMERIC MATERIALS

The nature of chemical bonds in polymeric materials is somewhat similar to

TABLE 6.2

RELATIVE CORROSION RESISTANCE OF SOME UNCOATED METALLIC MATERIALS

Material	*Industrial atmosphere*	*Fresh water*	*Sea water*	*Acids* H_2SO_4 *5-15% concentration*	*Alkalies 8%*	*Remarks*
Low carbon steel	1	1	1	1	5	Carbon steels and cast irons are used for their mechanical properties and low cost, they are not highly corrosion resistant. Their corrosion resistance is improved by Cr addition.
Galvanised steel	4	2	4	1	1	
Grey cast iron	4	1	1	1	4	
4–6% Cr steels	3	3	3	1	4	
18–8 stainless steel	5	5	4	2	5	Corrosion resistance is due to the formation of a protective film which is resistant to atmospheric and high temperature corrosion.
18–35 stainless steel	5	5	4	4	4	
Monel (70% Ni–30% Cu)	4	5	5	4	5	Nickel and Monel are particularly useful for handling alkalies and sea water.
Nickel	4	5	5	4	5	

Copper	4	4	4	3	3	Copper and its alloys are used for sea water resistance but acids tend to cause selective attack. Aluminium bronzes are the most resistant, particularly to acid conditions
Red brass (85% Cu–15%Zn)	4	3	4	3	1	
Aluminium bronze	4	4	4	3	3	
Nickel silver (65% Cu–18% Ni–17% Zn)	4	4	4	4	4	
Aluminium	4	2	1	3	1	Aluminium alloys can be anodised to improve their corrosion resistance.
Duralumin	3	1	1	2	1	

1 = Poor—rapid attack.
2 = Fair—temporary use.
3 = Good—reasonable service.
4 = Very good—reliable service.
5 = Excellent—unlimited service.

that in ceramics, insofar as all valence electrons participate in the bonding process and hence no free electrons exist in the structure. This means that electrochemical corrosion processes do not occur in polymers. However, other kinds of chemical processes affect the stability of polymers. Several organic fluids act as solvents for organic solids. Another problem is that almost all polymers absorb water to some extent. The absorbed water changes their strength and electrical properties and causes swelling and distortion of polymeric components. Table 6.3 lists the mositure absorption characteristics of some polymers. The table also gives the relative chemical stability of some polymers in different environments. Fluorocarbons (Teflon), are among the most chemically inert materials available to the engineer. Teflon is inert to all industrial chemicals and resists the attack of boiling *aqua regia*, fuming nitric acid or even hydrofluoric acid. Teflon is also stable in most organic solvents. It should be noted that the figures given in Table 6.3 are only comparative figures since the method of manufacture, surface finish or type of filler material can greatly affect the performance of polymers. For example, the water absorption of phenolics is 0·07% when filled with mica and 0·6% when filled with wood flour.

An effective way of stabilising polymeric materials against chemical attack involves reducing the rate of diffusion of the degradent into the polymer. This may be accomplished by developing a barrier at the polymer surface. For example, polyvinylfluoride films laminated to plastics are excellent surface erosion protectors. Structural modifications of the polymer can affect its resistance to chemical attack. For example, polycarbonates can be made more water resistant by increasing aromaticity in the system. However, modifications in the structure of a polymer to increase its resistance to chemicals can result in significant changes in its physical or mechanical properties.

6.5 SURFACE TREATMENTS FOR CORROSION PROTECTION

Although there is a wide variety of materials that can resist most industrial corrosive environments, it is not always practical to utilise such materials. Materials with the required bulk corrosion resistance may be expensive, as in the case of stainless steels or titanium, or may be mechanically weak, as in the case of pure aluminium. The corrosion problem can usually be solved by covering the corroding surface with a layer of a corrosion resistant material. The coating has to be selected such that it is compatible with the function and properties of the base material. Protective coatings can be

TABLE 6.3

RELATIVE CHEMICAL STABILITY OF SELECTED POLYMERIC MATERIALS

Material	*Water absorption after 24 h imersion (weight %)*	*Weak acids*	*Strong acids*	*Weak alkalies*	*Strong alkalies*	*Organic solvents*
Thermoplastics						
Fluorocarbons	0·00	5	5	5	5	5
Polyethylene	0·01–0·02	5	2	5	5	5
Polyvinylidene chloride	0·04–0·10	5	3	5	5	3
Vinyl chloride	0·45	5	3	5	3	2
Polycarbonate		5	3	5	3	2
Acrylics	0·03	3	3	5	3	1
Polyamides (Nylon)	1·50	3	1	5	3	3
Acetals		2	1	2	1	5
Polystyrene	0·04	3	1	3	1	1
Cellulose acetate	3·80	3	1	3	1	2
Thermosets						
Epoxy	0·10	5	3	5	5	3
Melamines	0·30	5	1	5	1	5
Silicones	0·15	5	3	3	2	3
Polyesters	0·01	3	1	2	1	2
Ureas	0·60	3	1	2	1	2
Phenolics	0·07–1·00	2	1	2	1	3

1 = Poor—rapid attack.
2 = Fair—temporary use.
3 = Good—reasonable service.
4 = Excellent—almost unlimited service.

metallic, ceramic or polymeric depending on the required function and service performance.

The chemical composition of a base metal can be changed by the surface diffusion of another metal at sufficiently high temperatures. Aluminising, or calorising, can be applied to iron or copper alloys by heating a component in a powder mixture of aluminium and aluminium oxide. The resulting surface layer can resist corrosion and oxidation up to about 1270 K. Chromising and sherardising are similar to aluminising, and involve heating the component in chromium and zinc powders respectively.

Hot-dip coatings involve dipping the component to be coated in a bath of molten metal. The process is called galvanising when coating with zinc, tinning when coating with tin and calorising when coating with aluminium. A possible disadvantage of this type of coating is that the base metal may be embrittled by the process, and costs may be high for some applications.

Corrosion protection can be provided through the formation of oxides on the metal surface by a process called conversion coating. Two processes are in commercial use: anodising is used for aluminium to give an oxide film which can be coloured; and phosphating is used for steel to give a phosphate film.

Electroplating is widely used for ferrous and non-ferrous base materials. The major plating metals are cadmium, zinc, copper, nickel and chromium. Cadmium and zinc are usually applied to steels for use in mildly corroding environments. Both metals provide anodic protection for steel and will be attacked preferentially, rather than the iron, if the plated layer is damaged. The two metals are also relatively cheap. Nickel and chromium plating are very versatile and provide good corrosion protection, excellent decorative surfaces and good wear resistance.

Vitreous enamels and ceramic coatings cover the surface to be protected with an adhering, hard and durable layer. Vitreous enamels usually consist of a borosilicate matrix in which crystalline opacifiers and pigments are suspended. Not all metals can be satisfactorily covered with vitreous enamels. Special compositions of iron and steel are especially designed and processed for enamelling. They are low in metalloids and may contain titanium or very low carbon. Vitreous enamels are hard and smooth with good abrasion resistance, but are susceptible to chipping. This disadvantage can be minimised by proper product design and careful selection of coating thickness. In general, the thinner the coating, the less the tendency to chip.

Another widely used class of finishes are organic coatings, which can be used on almost all types of materials. Protective coatings, with special reference to organic coatings, will be discussed in detail in Chapter 20.

6.6 COMBATING CORROSION WITH INHIBITORS AND CATHODIC PROTECTION

In many cases corrosion of metals in aqueous solutions may be reduced or eliminated by the addition of small quantities of soluble substances to the solution. These additives, called inhibitors, are either anodic or cathodic depending on which part of the corrosion cell is affected by their addition. Cathodic inhibitors are absorbed or plated out on cathodic areas, while anodic inhibitors precipitate with the anodic corrosion products, preventing further reaction by forming an adherent film on the metal surface. Such chemicals include phosphates, chromates and nitrites. Compared with cathodic inhibitors, anodic inhibitors can be more efficient, but are unsafe in that if insufficient inhibitor is available to prevent attack at every part of the metal, the unprotected areas are likely to suffer intensified attack.

Cathodic protection involves making the metal to be protected cathodic in the corrosion cell. The anode may be provided by: 1, a metal lower in the galvanic series, 2, a non-corroding material. For example, Ti or C, may be used as the anode by applying a potential difference to the system using a d.c. source, such as a battery, generator or rectifier. This method is most effective against nearly neutral solutions like salt water or soils. Underground steel pipelines can be protected by using the sacrificial corrosion of zinc or magnesium blocks buried near the pipe line and connected to it.

6.7 OXIDATION OF MATERIALS IN SERVICE

Many materials, metallic and non-metallic, combine with oxygen during service. In metals, oxidation often starts rapidly on exposure to oxidising environments and continues until an oxide film or scale is formed on the surface. After this stage, the rate of further oxidation depends on the soundness of the oxide film. If the oxide film is dense and impervious it provides protection against further oxidation; examples are Al_2O_3 on aluminium and Cr_2O_3 on chromium. In most metals, the oxide film tends to be in compression, since the volume of the oxide is greater than the volume of metal consumed in its formation. Under these conditions, the metal oxidises at a decreasing rate with time, but at an increasing rate with a rise in temperature.

The growth of an oxide film can occur in a number of different ways; each way can be described by a growth formula. Parabolic growth is observed when the rate of diffusion of metal and oxygen ions in the film governs its growth. At a given temperature parabolic growth can be described by the formula:

$$T^2 = K_1 t \tag{6.1}$$

where T is the film thickness, t is time and K_1 is a constant for the metal. At room temperature, metals like iron, nickel and copper oxidise in a parabolic manner. Linear growth is observed when the rate of oxidation at a given temperature is constant as given by the formula:

$$T = K_2 t \tag{6.2}$$

where K_2 is a constant. Metals like sodium and potassium behave in this manner, since their oxide films are porous and allow continuous access of oxygen to the metal surface. Logarithmic growth describes the case of protective oxide films that practically stop growing after a critical limiting thickness is reached, as in the case of aluminium, chromium and zinc. For a given temperature the oxide thickness (T) is given by the formula:

$$T = K_3 \log (ct + 1) \tag{6.3}$$

where K_3 and c are material constants.

In many cases the composition and characteristics of oxide films can be changed by alloy addition to the base metal, e.g. the addition of chromium to iron modifies the normally porous iron oxide layer and makes it impervious and protective. Chromium, aluminium and silicon are usually added to modify the oxide films of iron, nickel and cobalt. Figure 6.1 illustrates the effect of Cr, Al, and Si additions on the oxidation resistance of steel.

In some heat resisting alloys, the addition of alloying elements to improve oxidation resistance can have undesirable effects on their mechanical performance. Under these conditions, the alloys can be protected against oxidation by applying ceramic coatings. These coatings are similar to vitreous enamels, except that they have better high temperature resistance. The most temperature resistant ceramic coatings consist of pure metal oxides such as alumina and zirconium oxide. They are applied by flame or plasma spraying. The molten ceramic particles impinge on the preheated base material and cover it with a coating of uniform thickness. Many materials can be coated in this way, for example, ferrous metals, graphite and glass.

Most plastics and rubbers oxidise in the presence of oxygen. Rubbers are especially susceptible to oxidation and the process is called ageing. The

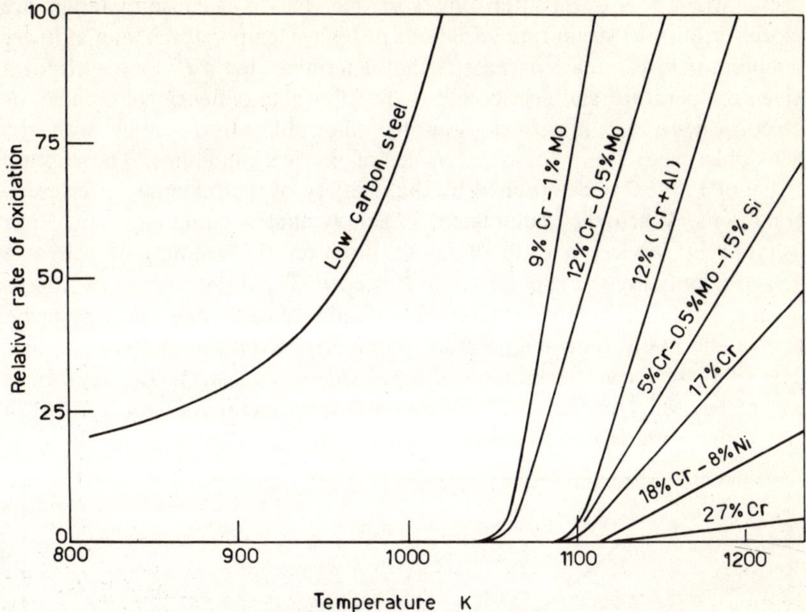

FIG. 6.1. Schematic representation of the effect of Cr, Al and Si additions to steel on its oxidation in air.

reaction of oxygen with rubber initially reduces elasticity and increases hardness. This is because oxygen diffuses into the structure and provides additional cross-linking. As ageing proceeds, the rubber degrades, and eventually loses most of its strength. The rate of ageing depends on temperature, type of atmosphere and material composition and method of manufacture.

6.8 STABILITY OF METALLIC MATERIALS AT ELEVATED TEMPERATURES

The effect of service ernvironment on material performance at elevated temperature can be divided into three main categories: 1, chemical effects like oxidation; 2, mechanical effects like creep and stress rupture; and 3, microstructural effects like grain growth and overageing. Oxidation of metallic materials was discussed in Section 6.7, while mechanical and microstructural effects will be discussed here.

Generally, materials become substantially weaker as their service temperature increases. The loss of strength is a function of time, so that useful strength is lower after longer service. This is because materials are more sensitive to strain-rate variations at higher temperatures than at lower temperatures. The main parameter that determines the useful strength for a given temperature and service life is the allowable dimensional change, or creep. Figure 6.2 illustrates how the allowable stress varies with the allowable creep-strain for a given set of service conditions. The rupture curve of Fig. 6.2 is determined by the ductility of the material under creep conditions. When the requirement is simply that a component must not fracture in service, and there is no limit on the amount of tolerable deformation, only the rupture curve is needed. The different rupture curves obtained at different temperatures can be reduced to a single curve by using one of the many time–temperature parameters available in the literature. One of these parameters that has been widely used is the Larson–Miller parameter, which expresses creep stress (σ) as a function of temperature (T) and rupture life (t_r) as follows:

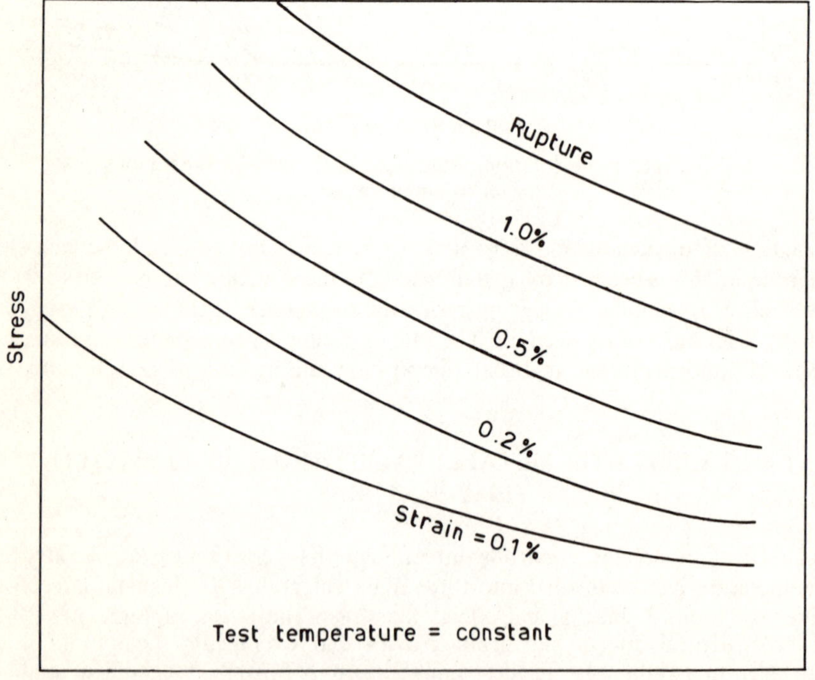

FIG. 6.2. Schematic representation of the stress–creep-strain data at constant temperature.

$$\sigma = f[T(c + \log t_r)] \tag{6.4}$$

where c is a constant.

Stress relaxation is closely associated with creep at high temperatures. It takes place in components subjected to constant strain at high temperatures. The stress required to maintain the component under strain gradually diminishes with time. Practical examples where stress relaxation is important are: bolts, studs, flanges and springs. In a bolted joint if the stress in the bolts is relaxed below a certain value leaks may develop.

Unlike room temperature service conditions, high temperature service can change the structure of the material and consequently the mechanical behaviour. Many of the strengthening mechanisms that were discussed in Section 1.3 become ineffective at high temperatures. Thus, materials which depend on their fine grains for strengthening may lose this advantage by grain growth; structures which have been precipitation hardened to peak values may overage; steels which have been quenched and tempered may overtemper; and materials which have been strain-hardened by cold working may recover or anneal. This indicates that treatments which improve the room temperature mechanical properties or even the high temperature, short term strength do not necessarily improve the creep resistance. Generally, non-equilibrium structures change during long term, high temperature service and this leads to lower creep strength.

Dispersion hardened alloys, described in Chapter 5, are similar in some respects to precipitation hardened alloys, but are more stable and give better service performance under creep conditions. Examples of dispersion hardened alloys are SAP ($Al–Al_2O_3$) and TD nickel ($Ni + ThO_2$). In both cases the dispersed phase does not dissolve back into the matrix, and does not coarsen at a fast rate, which accounts for its stability at high temperatures. Another example is that the use of plain carbon steels for high temperature applications is limited by the instability of iron carbides. The addition of alloying elements like tungsten, chromium, vanadium and molybdenum stabilises the carbides and markedly reduces the rate of their coalescence at elevated temperatures; the resulting material is known as high speed steel.

6.9 STABILITY OF NON-METALLIC MATERIALS AT ELEVATED TEMPERATURES

As oxide ceramic materials are very stable at high temperatures, their use for prolonged periods of time at such temperatures presents no problem. However, if service conditions involve rapid fluctuations in temperature, the material may suffer thermal damage due to thermal shock (spalling). A higher thermal expansion coefficient of the material, a higher elastic modulus and a lower thermal diffusivity will result in greater thermal damage and lower thermal shock resistance.

Polymers are also prone to thermal damage which is caused by thermally activated depolymerisation mechanisms. A reduction of the degree of polymerisation affects the mechanical properties and the performance of the polymer. Degradation can also occur by chemical decomposition into non-monomeric molecules such as H_2O, CH_2O, CH_4 and CH_3OH. Charring of thermosetting plastics represents the last stages of thermal damage.

6.10 RADIATION EFFECTS

Engineering materials may be subjected to one of the many sources of radiation during the course of their service. Unlike elevated temperatures, which energise all the atoms within a solid, radiation processes may focus relatively large amounts of energy locally. High energy radiation sources can be divided into two major classes, electromagnetic radiation and particle radiation. In the class of electromagnetic radiation, gamma rays have shorter wavelength than X-rays and consequently have more energy and are more penetrating. The energy of atomic and nuclear particles is much higher than that of electromagnetic radiation, and depends on the mass and velocity of the particles. Important types of particle radiation are generated in nuclear fission and fusion reactions. The particles involved may be electrons (beta rays), fast charged particles from an atomic nucleus (protons or deuterons) and fast uncharged particles (neutrons).

X-rays, gamma rays and beta rays usually affect the electrons in the structures of solids and cause ionisation. In metallic materials, ionisation causes almost no damage due to the presence of free electrons which quickly neutralise the electronic disturbance. However, ionisation can cause permanent changes in ionic and covalent structures such as those in ceramics, glasses and polymers. Disturbance of the chemical bonds in these materials can change their characteristics.

Particle radiation interacts primarily with nuclei, knocking complete atoms out of their equilibrium position. The amount of radiation damage depends on the energy of the radiation particles, and on the material subjected to them. Materials with a large cross section for a particular type of radiation will suffer more damage, since they capture more particles within their lattice.

Radiation damage in metallic materials usually increases their hardness and reduces their ductility in a similar way to cold working. Similarly, annealing will soften the material back to its original condition. Despite these similarities however, the structural mechanisms are basically different, because point defects are mainly responsible for radiation damage, while linear defects are mainly responsible for strain hardening. Table 6.4 illustrates the effects of radiation on the properties of some metallic materials.

TABLE 6.4

EFFECT OF RADIATION ON THE MECHANICAL PROPERTIES OF SOME METALLIC MATERIALS (NEUTRON FLUX OF ABOUT 10^{20}N/CM^2)

Material	*Tensile strength* (*MN/m*^{2})		*Yield strength* (*MN/m*^{2})		*Elongation* (%)	
	before	*after*	*before*	*after*	*before*	*after*
Pure iron	252	259	126	217	42	6
C1035 steel	518	714	329	672	34	11
302 stainless steel	686	805	259	679	63	52
1100 aluminium alloy	119	182	49	119	40	23
2024 aluminium alloy	504	595	315	462	26	24
Inconel X	1 218	1 246	826	1 183	29	10

In polymeric materials, commercial use has been made of the effect of ultraviolet radiation because it causes branching in polyethylene, which increases its stability at the temperature of boiling water. Neutron radiation, however, may cause degradation of polymers which adversely affects their properties.

In selecting a material for an application where a certain type of radiation is present, there are two alternatives available:

1. Select a material that will not suffer structural damage, i.e. that has a small cross section for the radiation.
2. Shield the material from the radiation. The chosen alternative will depend on the service requirements and the design configuration.

BIBLIOGRAPHY

1. G.T. Bakhvalov and A.V. Turkovskaya, *Corrosion and Protection of Metals*, Pergamon (Oxford, New York), 1965.
2. H.R. Clauser, *Industrial and Engineering Materials*, McGraw-Hill (New York, Maidenhead England), 1975.
3. D.R. Gabe, *Principles of Metal Surface Treatment and Protection*, Pergamon (Oxford, New York), 1972.
4. Z.D. Jastrzębski, *The Nature and Properties of Engineering Materials*, 2nd Edn, John Wiley and Sons (New York), 1976.
5. B. Kepelman, *Materials for Nuclear Reactors*, McGraw-Hill (New York, Maidenhead England), 1959.
6. G. Koves, *Materials for Structural and Mechanical Functions*, Hayden (New York), 1970.
7. D. Stewart and D.S. Tulloch, *Principles of Corrosion and Protection*, Macmillan (New York), 1968.
8. L.H. Van Vlack, *Materials Science for Engineers*, Addison-Wesley (Reading Mass., London, Amsterdam), 1970.

7

Failure in Service

7.1 INTRODUCTION

The service behaviour of a material is governed not only by its inherent properties but also by the stress system acting on it and the environment in which it is operating. Components fail in service when the incorrect material has been selected for their manufacture, or when the service conditions are more severe than those anticipated by the designer. Manufacturing defects also account for a considerable proportion of cases of failure.

Anticipating the different ways in which a product could fail is an important factor that should be considered, when selecting a material or a manufacturing process for a given application. The possibility of failure of a component can be analysed by studying on the job material characteristics, the stresses and other environmental parameters that will be acting on the component, and the possible manufacturing defects that can lead to failure. This technique is called failure analysis and can be carried out in the following ways:

1. By an environment profile that provides a description of the expected service conditions. These include operating temperature and atmosphere, radiation, presence of contaminants and corrosive media, other materials in contact with the component and the possibility of galvanic corrosion, and lubrication conditions.
2. By fabrication and process flow diagrams that provide an account of the effect of the various stages of production on the material properties, and of the possibility of quality control. Certain processes can lead to undesirable directional properties, internal stresses, cracking, or structural damage, which could lead to unsatisfactory component performance and premature failure in service.
3. By failure models that describe all the possible types of failure and the

conditions that can lead to them. Failure due to chemical causes occurs when corrosion, chemical reaction or oxidation are so excessive that it becomes hazardous for the component to remain in service. Electrical failures occur in insulating materials when the applied voltage exceeds the breakdown voltage of the material, or when flash-over takes place. Mechanical failures are more varied and represent a serious threat to all load bearing components and will be discussed in more detail in this chapter.

7.2 TYPES OF MECHANICAL FAILURE

Generally, a component can be considered to have failed when it does not perform its intended function with the required efficiency. The general types of mechanical failure encountered in practice are:

1. Yielding of the component material under static loading. Yielding causes permanent deformation which could result in misalignment or hindrance to mechanical movement.
2. Buckling, which takes place in slender columns when they are subjected to compressive loading, or in thin-walled tubes when subjected to torsional loading.
3. Creep failure, which takes place when the creep strain exceeds allowable tolerances and causes interference of parts. In extreme cases failure can take place through rupture of the component subjected to creep. In bolted joints and similar applications failure can take place when the initial stressing has relaxed below allowable limits, so that the joints become loose or leakage occurs.
4. Failure due to excessive wear, which can take place in components where relative motion is involved. Excessive wear can result in unacceptable play in bearings and loss of accuracy of movement. Other types of wear failure are galling and seizure of parts.
5. Failure by fracture due to static overload. This type of failure can be considered as an advanced stage of failure by yielding. Fracture can be either ductile or brittle.
6. Failure by fatigue fracture due to overstressing, material defects or stress raisers. Fatigue fractures usually take place suddenly without apparent visual signs.
7. Failure due to the combined effect of stresses and corrosion, which usually takes place by fracture due to cracks starting at stress

concentration points, for example, caustic cracking around rivet holes in boilers.

8. Fracture due to impact loading, which usually takes place by cleavage in brittle materials, for example in steels below brittle–ductile transition temperature.

Of the above types of mechanical failure, the first four do not usually involve actual fracture, and the component is considered to have failed when its performance is below acceptable levels. On the other hand, the latter four types involve actual fracture of the component, and this could lead to unplanned load transfer to other components and perhaps other failures. This can be avoided by careful design and the selection of the appropriate factor of safety as discussed in the following section.

7.3 FACTOR OF SAFETY

In designing a component for a given application two types of service conditions have to be specified: 1, normal working conditions which the component has to endure during its intended service life; and 2, limit working conditions, such as overloading, which the component is only intended to endure on exceptional occasions, and which if repeated frequently could cause premature failure of the component. In a mechanically loaded component, the stress levels corresponding to both normal and limit working conditions can be determined from a duty cycle. The normal duty cycle for an airframe, for example, includes towing and ground handling, engine run, take-off, climb, normal gust loadings at different altitudes, kinetic and solar heating, descent and normal landing. Limit conditions can be encountered in abnormally high gust loadings or emergency landings. Analyses of the different loading conditions in the duty cycle lead to determination of the maximum load that will act on the component. This maximum load value can be used to determine the maximum stress, or damaging stress, which if exceeded would render the component unfit for service before the end of its normal expected life.

The life of a component can be estimated according to safe-life or fail-safe criteria. The safe-life criterion can be applied to components in which an undetected crack or other defect could lead to catastrophic structural failure, and a life limitation must therefore be imposed on their use. The fail-safe criterion can be applied to structures in which there is sufficient tolerance of a failure to permit continuous service until discovered by routine inspection procedure, or by obvious functional deficiencies. The majority of engineering components can be designed according to the fail-safe criterion. Even a critical component can be designed according to the fail-safe criterion if failure is detectable by the maintenance programme, which must define both the timing and the methods of inspection to be applied. Redistributing the load into sufficiently robust adjacent components if failure occurs is an added safety precaution. If the use of a safe-life component is unavoidable, its safe service life must be established by testing, and its replacement life calculated by applying an appropriate factor of safety.

The factor of safety can be taken as the ratio of the damaging stress to the design stress. If the exact service loading conditions and the exact performance of the component can be determined, a factor of safety approaching unity can be employed. In practice, the designer usually uses factors of safety from 1·2 to 20. Higher values of factor of safety are used when the material properties are not homogeneous, with the possibility of finding inclusions and porosity. Inaccuracies in determining the stress distribution in the component, and the presence of internal stresses and stress concentrations, call for a higher factor of safety. Common values of the factor of safety range from 1·5 to 10.

For ductile metallic materials, under static loading, the yield strength can be taken as the damaging stress; but for brittle materials the ultimate strength is usually taken as the damaging stress. Under fatigue conditions, the endurance limit is the damaging stress.

Unless exceptionally severe overloads are encountered, components do not usually fail through tension or shear. On the other hand fatigue failures account for about 80% of machine part failures, because the stresses required to produce them are much lower than the static strength of the material, as will be discussed in the following section. Failures due to lack of impact toughness of materials also occur in certain mechanical applications, and will be discussed later in this chapter.

7.4 FATIGUE FAILURE

A component can fail without warning under repeated applications of stresses that it could support indefinitely if the stresses were static. Fatigue failures are generally initiated by one or more cracks starting at some discontinuity in the material or at other stress concentration locations. After initiation, the fatigue crack usually spreads gradually through the component. The duration of the crack propagation stage depends on the loading conditions, and involves very little change in the overall dimensions of the component. At relatively low stresses, the propagation stage can take as long as 80 or 90% of the fatigue life. Complete fracture of the component occurs suddenly, when the remaining uncracked material is insufficient to support the applied load. The general appearance of fatigue fractures is brittle, even in the case of ductile materials.

The fatigue resistance of materials generally increases with the increase in their ultimate tensile strength, and in the absence of fatigue data the endurance ratio can be used to get a rough estimate of fatigue strength. The endurance ratio is defined as the ratio of fatigue strength to ultimate tensile strength. This ratio is approximately constant for a given material. Table 7.1 compares the tensile strength and fatigue strength for selected ferrous metallic materials where the endurance ratio varies between 0·4 and 0·6. Table 7.2 compares the tensile and fatigue strengths for selected non-ferrous and polymeric materials. It should be noted that the fatigue strength values are subject to large variation, and depend on the type of test and on whether the strength is taken at 10^6, 10^7 or 10^8 cycles to fracture. The size of the test specimen and its surface finish also affect the value for fatigue strength.

Fatigue cracks usually start at the surface of the component, although in some cases they may be initiated within a material, especially at high stress levels. Surface cracks, however, are more common because bending or torsion will cause the highest stresses to occur at the outer fibres. Typical surface crack initiating spots are at keyways, thread roots, sharp corners, machining marks and weld faults. As a result, the fatigue strength of a component depends a great deal on the condition of its surface, as illustrated in Table 7.3 for a variety of steels. The table also illustrates, the detrimental effects of corrosive media on fatigue strength. Because of the drastic effect of corrosion on fatigue behaviour, it is usually found that the corrosion resistance, rather than the fatigue strength, determines the fatigue life of a component.

Components which cannot be polished to improve their fatigue behaviour can be treated by shot-peening, surface hardened by induction or by flame, nitrided or carburised. These treatments increase the surface hardness and introduce surface compressive stresses that increase the fatigue strength.

The above discussion shows that the presence of stress concentrations is one of the major factors that decide the fatigue behaviour of a component. A measure of the degree of notch sensitivity of the material is usually given by the parameter q:

$$q = \frac{K_f - 1}{K_t - 1}$$

TABLE 7.1

COMPARISON OF STATIC AND FATIGUE STRENGTHS OF FERROUS ALLOYS

Material	*Treatment*	*Ultimate tensile strength (MN/m^2)*	*Endurance limit (MN/m^2)*	*Endurance ratio*
Carbon steels				
AISI 1010	Normalised	364	186	0·46
AISI 1025	Normalised	441	182	0·41
AISI 1035	Normalised	539	238	0·44
AISI 1045	Normalised	630	273	0·43
AISI 1060	Normalised	735	315	0·43
AISI 1060	Oil quenched and tempered at 723 K	1 295	574	0·44
ASTM A27, grade 60–30	As cast	420	210	0·50
SAE automotive grade 0050B	As cast	700	315	0·45

Alloy steels				
AISI 3325	Oil quenched and tempered at 923 K	854	469	0·55
AISI 4340	Oil quenched and tempered at 873 K	952	532	0·56
AISI 8640	Oil quenched and tempered at 898 K	875	476	0·54
AISI 9314	Oil quenched and tempered at 873 K	812	476	0·59
ASTM A352 grade LC1	As cast	455	224	0·49
ASTM A148 grade 90–60	As cast	630	294	0·47
Stainless steels				
AISI 302	Annealed	560	238	0·43
AISI 316	Annealed	560	245	0·44
AISI 410	Quenched and tempered at 1023 K	518	217	0·42
AISI 431	Quenched and tempered at 923 K	798	336	0·42
Cast irons				
Grey C.I. ASTM 20	As Cast	140 (min)	70	0·50
Grey C.I. ASTM 30	As cast	210 (min)	102	0·49
Grey C.I. ASTM 60	As cast	420	168	0·40

TABLE 7.2

COMPARISON OF STATIC AND FATIGUE STRENGTHS OF NON-FERROUS AND POLYMERIC MATERIALS

Material		*Treatment*	*Tensile strength (MN/m^2)*	*Endurance limit (MN/m^2)*	*Endurance ratio*
Aluminium alloys					
Wrought	2011	T8	413	245	0·59
	2024	annealed	189	91	0·48
	6061	annealed	126	63	0·50
		T6	315	98	0·3
	6063	T6	245	70	0·29
	7075	T6	581	161	0·28
Cast	214	as cast	175	49	0·28
	380	die-cast	336	140	0·42
	356	T6	266	91	0·34
Copper alloys					
Phosphor bronze		annealed	315	189	0·6
(5% Sn)		hard drawn	602	217	0·36
Silicon bronze		annealed	259	105	0·41
(1·5% Si)		half hard	336	175	0·52
Aluminium bronze (9·5% Al)		quarter hard	581	206	0·35
Cupro–nickel		hard drawn	511	228	0·45
(30% Ni)		cold drawn	672	273	0·41
Monel (63% Ni)					
Superalloys					
Incoloy 901		tested at 923 K	980	364	0·37
Udimet 700		tested at 1073 K	910	343	0·38
Waspaloy		tested at 1073 K	700	280	0·4
Haynes alloy 25		tested at 1073 K	350	217	0·62
Reinforced polymers					
Glass fibre reinforced polyester			123	84	0·68
Glass fibre epoxy laminate			350–490	196	0·56–0·4

where K_f is the ratio of the fatigue strength, in the absence of stress concentrations, to the fatigue strength, with stress concentrations, and K_t is the stress concentration factor which represents the severity of the notch and is given by the ratio of maximum local stress at the notch to average stress.

TABLE 7.3

EFFECT OF SURFACE CONDITION ON THE FATIGUE STRENGTH OF STEELS*

Ultimate tensile strength (MN/m^2)	*Measured endurance strength as a percentage of maximum endurance*						
	Mirror polish	*Poli-shed*	*Machi-ned*	*0·1mm notch*	*Hot work-ed surface*	*Under fresh water*	*Under salt water*
280	100	95	93	87	82	72	52
560	100	92	88	77	63	53	36
840	100	90	84	66	47	37	25
1 120	100	88	78	55	37	25	17
1 400	100	88	72	44	30	19	12
1 540	100	88	69	39	30	19	14

*In this table steels are classified according to their ultimate tensile strength.

The value of q can be considered as a measure of the degree of agreement beteen K_f and K_t. Thus, as q increases from 0·0 to 1·0, the material becomes more sensitive to the presence of stress concentrations. Generally, increasing the tensile strength of the material makes it more notch sensitive and increases q. The value of q is also dependent on component size, and it increases as size increases, and so stress-raisers are more dangerous in large masses.

In designing a component for fatigue resistance, an important step is to analyse the stresses, taking into consideration both the alternating and static loads. The magnitude and nature of the static stresses could exert a large influence on the fatigue performance of the component. The stress analysis stage is followed by the testing of specimen components under conditions as close as possible to the expected service conditions. The results of these tests show the points of weakness, and help in modifying the design to arrive at the optimum performance.

7.5 SELECTION OF MATERIALS AND PROCESSES FOR RESISTANCE TO FATIGUE

Besides the factors that have been discussed, the fatigue performance of a component is largely influenced by the type of material and the processes used in making it. Steels are the most widely used structural materials for fatigue applications as they offer high fatigue strength and good processability at a relatively low cost. Steels have the unique characteristic of exhibiting an endurance limit which enables them to perform indefinitely, without failure, if the applied stresses do not exceed this limit. As shown in Table 7.1 the endurance limit is roughly equal to 0·4 to 0·6 of the ultimate tensile strength of steels up to UTS of about 1100 MN/m^2. With stronger steels, the scatter in fatigue strength becomes wide and the endurance ratio can become less than 0·30.

The optimum structure for fatigue resistance is tempered martensite, since it provides maximum homogeneity. Steels with high hardenability give high strength with relatively mild quenching and, hence, low residual stresses, which is desirable in fatigue applications. Normalised structures give better fatigue resistance than pearlitic structures obtained by annealing. Inclusions in steel are harmful as they represent discontinuities in the structure that could act as initiation sites for fatigue cracking. Therefore, free machining steels should not be used for fatigue applications. However, if machinability considerations make it essential to select a free machining grade, the leaded steels are preferable to those containing sulphur or phosphorus. This is because the rounded lead particles give rise to less structural stress concentrations than the other angular and elongated inclusions. By the same token, cast steels and cast irons are not recommended for critical fatigue applications. In rolled steels, the fatigue strength is subject to the same directionality as the static properties.

The great majority of non-ferrous alloys do not have an endurance limit, as in the case of ferrous alloys, and their fatigue strength is usually taken at a given number of loading cycles, usually 10^7 or 10^8 cycles. Aluminium alloys usually combine corrosion resistance, light weight and reasonable fatigue resistance. The endurance ratio of aluminium alloys is more variable than that of steels (Table 7.2), but an average value can be taken as 0·35. Generally, the endurance ratio is lower for as-cast structures and precipitation hardened alloys. Fine-grained inclusion free alloys are most suited for fatigue applications.

Copper alloys, like most other non-ferrous alloys, have no endurance limit and the endurance ratio can vary over a wide range, as shown in Table 7.2. Some bronzes and Cu–Ni alloys have good fatigue strengths which can be improved by cold working.

Plain polymeric materials are usually not suitable for high fatigue load applications but reinforced grades can have high fatigue strength/weight ratios. Table 7.2 gives the fatigue strength of some glass-fibre-reinforced plastics, but fatigue data are difficult to find and show considerable inconsistency.

As would be expected, crystalline ceramic materials do not have significant fatigue strength in view of their brittleness and inhomogeneity. However, glass and ceramic fibres can be effectively utilised as reinforcing agents for ductile but soft materials.

7.6 BRITTLE FRACTURE

Brittle fracture may be defined as a form of fracture that occurs suddenly under a load which is not sufficient to cause general yielding across the whole of the fractured surface. Most brittle fractures initiate from some form of notch, giving rise to a stress concentration. If the local yield at the tip of the notch is insufficient to spread the stress over a large area a brittle fracture may be initiated. Once started, the brittle fracture will run at high speed, reaching 1200 m/s in steel, until total failure occurs, or until it runs into conditions favourable for its arrest. The risk of occurrence of brittle fracture depends on the notch toughness of the material under a given set of service conditions.

A characteristic feature of brittle fracture surfaces is the chevron pattern, which consists of a system of ridges curving outwards from the centre line of the plate. These ridges, or chevrons, may be regarded as arrows with their points on the centre line and invariably pointing towards the origin of the fracture, so providing an indication of its propagation pattern. This feature is useful in the analysis of service failures.

The temperature at which the component is working is one of the most important factors that influence the nature of fracture. Brittle fractures are usually associated with low temperature, and in some steels conditions may exist where a difference of a few degrees, even within the range of atmospheric temperatures, may determine the difference between ductile and brittle behaviour. This sharp ductile–brittle transition is only observed in BCC and CPH metallic materials and not in FCC materials. Figure 7.1 illustrates the variation with temperature of the energy required to fracture steel specimens in a Charpy V-notch test. This energy can be taken as a measure of the toughness of the material.

The speed of loading also influences the nature of fracture. Materials which behave normally under slowly applied loads may behave in a brittle manner when subjected to sudden applications of load, such as shock or impact. The shape of the notch affects the state of stress and the value for stress concentration, which can affect the initiation of brittle fracture. The

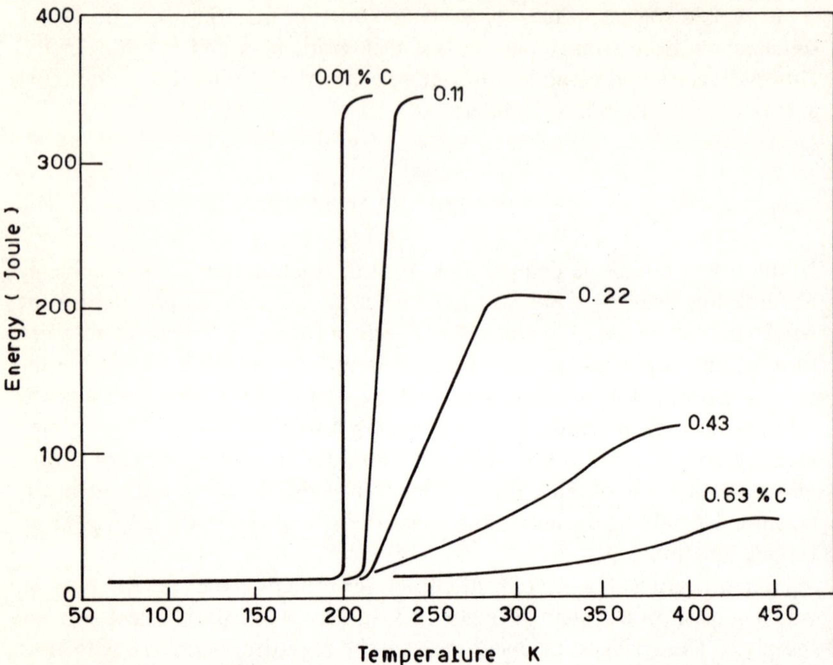

FIG. 7.1. Effect of temperature on the energy required to fracture Charpy V-notch specimens of plain carbon steel in the normalised and annealed condition.

notches in a component can be due to shape changes, processing defects or corrosion attack.

the chemical composition and structure of the material play an important role in controlling its fracture behaviour. Brittleness is favoured by large grain size and by brittle inclusions. Figure 7.1 shows that steels become less tough with increasing carbon content. Experience has shown that a large proportion of brittle fractures originate from welds or their vicinity. This can be caused by the residual stresses generated by the welding process, reduction of the toughness of the heat affected zone or by defects in the weld metal.

The above discussion shows that the response of materials to loading is affected by several mechanical and metallurgical parameters, and it would be convenient if a single measurable parameter could be used to anticipate the behaviour of a component in service. To this end, industrial practice has moved towards standardising on the Charpy V-notch impact test as a measure of the notch toughness of materials. The variation of toughness with test temperature has already been shown in Fig. 7.1. The curves of Fig. 7.1 can be used to anticipate the behaviour of materials in service. The temperature at which the material behaviour changes from ductile to brittle is called the ductile–brittle transition temperature (*Tc*), and may be taken as the temperature at which the fractured surfaces exhibit 50% brittle fracture appearance. In V-notch Charpy experiments the transition temperature can be set at a level of 15 ft/lbf (20·3J) or at 1% lateral contraction at the notch. The transition temperature based on fracture appearance always occurs at a higher temperature than if based on a ductility or energy criterion. Therefore the fracture appearance criterion is more conservative.

In applying the Charpy V-notch results to industrial situations it should be borne in mind that the shock conditions encountered in the test may be too drastic. Many industrial components operate successfully in extreme cold without special consideration for notch toughness values or transition temperature. However, where stress concentration and rate of strain are high and service temperatures are low, materials with low transition temperatures should be selected. Under these conditions the following design and fabrication considerations should be applied:

1. Abrupt changes in section should be avoided and thickness should be kept to a minimum.
2. Welds should be located clear of stress concentrations and of each other, and should be easily accessible for inspection.
3. Whenever possible, welded components should be designed on a fail-safe basis.

7.7 SELECTION OF MATERIALS AND PROCESSES FOR NOTCH TOUGHNESS

As in many load bearing applications, metallic materials are generally used where high notch toughness is required. Some polymeric materials can also be used in applications where special additional characteristics are required, e.g. high dielectric properties, light weight and light transparency. Ceramics are generally too brittle to be used for most applications involving impact loading.

The major criterion for measuring the impact performance of steels is the ductile–brittle transition temperature (*Tc*), which should be lower than the minimum expected service temperature. Generally, *Tc* is lower for lower carbon content steels (Fig. 7.1), and for optimum performance a low carbon steel with uniform, fine pearlitic structure is preferable. A thoroughly deoxidised steel grade has fewer non-metallic inclusions and gives better performance. Table 7.4 presents some data on the notch toughness of some steel grades. The ranges given can be expected in commercial materials as a result of variations in composition, grain size and structure. The use of high-carbon steels for impact applications is not recommended, and if high static strength is required austenitic stainless steels can be used. Ferritic and martensitic stainless steels are not recommended.

The major use for FCC non-ferous alloys in notch toughness applications is at cryogenic temperatures where most BCC steels become too brittle to use. Several aluminium, titanium, copper and nickel base alloys are available for cryogenic application. The selection of these materials will be discussed in detail in Chapter 19.

Some polymeric materials are available in high impact grades but even these have much lower impact strengths than most metallic materials. Table 7.5 lists the properties of selected polymeric materials. The given values are obtained from a special Izod test and are not directly comparable with the impact toughness values given in Table 7.4. In view of the high cost of polymers in comparison with carbon steels, it is possible that the selection of an impact grade of plastics may impose a cost penalty that should be justified by other advantages, such as light weight, dielectric properties, processability, corrosion resistance and appearance.

TABLE 7.4

CHARPY V-NOTCH IMPACT TOUGHNESS OF SELECTED STEELS

Steel grade	*Treatment*	*Impact energy at 20°C (Joule)*
AISI 1018	hot rolled	20–75
AISI 1025	hot rolled	2·7–48
AISI 1030	hot rolled	2·7–21
AISI 8617	heat treated to RC 30	100–136
	heat treated to RC 35	65–95
AISI 8645	heat treated to RC 40	38–55
	heat treated to RC 50	19–25
AISI 6150	quenched and tempered to RC 40	20–30
	austempered to RC 40	13–30
301 stainless steel	cold rolled	108–163
347 stainless steel	cold rolled	330–360
431 stainless steel	annealed	47–68

TABLE 7.5

IMPACT TOUGHNESS OF SOME POLYMERIC MATERIALS

Material	*Izod impact energy of notched specimens (Joule)*
ABS (high impact grade)	8–14
Acetals	1·6–2·5
Acrylics	0·5–3·2
Epoxies (glass filled)	up to 40
Ethyl cellulose	4·75–8·2
Fluorocarbons	3·4–5·5
Melamines	0·4–1·1
Phenolics (glass fibre filled)	13–40
Polyesters(glass filled)	up to 10
Polyethylenes	2·0–27
Polycarbonates	5·4–19
Silicones (glass filled)	up to 27
Ureas	0·3–0·5
Vinyl, chloride–acetate	0·5–2·7

7.8 PRODUCT RELIABILITY

Reliability of a product in service is related to the probability that it will operate for a predetermined period of time without failure under specified usage conditions. Generally, there are four types of product failure in service:

1. Early failure due to faulty material or manufacturing errors which have not been detected by quality control. Early failures can be minimised by exercising sound quality control procedures during the different production steps.
2. Random hazard failures which take place during the product's service life due to chance. Accidental clogging of lubrication passages or a surge of electric power supply can cause this type of product failure.
3. Excessive overloading of the product beyond its safe capacity, which can cause its failure.
4. Wear-out failures, which are natural in all products when they reach the end of their intended service life. This type of failure is usually progressive and the product gradually loses its functional efficiency. Wear-out can be postponed by proper preventive maintenance.

The probability of product failure due to early and excessive overloading is usually higher in the early stages of the expected product life. Random hazard failures have an equal probability of taking place at any time, while wear-out failures occur more frequently near the end of the expected product life. When a product is an assembly of several components and each component has its own pattern of failure probability, then the reliability of the product (R_p) is given as:

$$R_p = R_1 \times R_2 \times R_3 \times \ldots$$

where R_1, $R_2 \ldots$ etc. are the reliability of components 1, 2, ... etc. The reliability of the product can never exceed the reliability of the components involed in making it. This illustrates the importance of quality control at all stages of production. Product life can be short for cheap non-critical products, but it must be long for large expensive machines. Early failures should be eliminated for aircraft and similar products.

7.9 QUALITY CONTROL

Quality control may be defined as all those activities that affect the quality of products at the different stages of their manufacture. Therefore, quality control includes the processes of testing and measuring performance, with the aim of accepting, rejecting, adjusting or reworking a component or a product. Quality control may involve the inspection of every component in the product, i.e. 100% inspection, or the inspection of random samples of each component. Another technique is to test partial component assemblies, or even just the completed product, using random or 100% inspection. It should be noted that 100% inspection does not mean complete elimination of defective parts. Ordinary inspection equipment has its own efficiency, and usually the human factor is also present. A 100% inspection is usually less than 98% efficient.

The fact that no result can be absolutely assured and nothing is perfect in industrial production has led to the evolution of statistical quality control. In this technique observable features of a product or process are measured. The percentage of measurements falling within limits can be determined by the sampling method, and gives the same answer as if the product were 100% inspected. The size of sample is determined by the nature of the product and the standard of quality required. The standard deviation is a useful measure of the extent to which an item may deviate from the expected mean. To employ the standard deviation, confidence limits must be adopted to specify the percentage probability that the product will fall between the upper and lower limits. The standard deviation (S) can be defined as:

$$S = \sqrt{\frac{1}{(n-1)} \sum_{i=1}^{n} (x_i - \bar{x})^2}$$

where $\bar{x}$ is the mean value of the set of numbers (x_i), and n is the number of items in the set.

$$\bar{x} = \frac{1}{n} \sum_{i=1}^{n} x_i.$$

The accuracy with which the mean, and the standard deviation, of the sample represent those of the population, increases with the size of the sample of data taken.

The adoption of statistical quality control techniques has the advantage of reducing inspection costs while at the same time ensuring high quality. Production troubles are also anticipated which leads to more uniform products with less rejects, scrap and rework.

BIBLIOGRAPHY

1. G.M. Boyd, Ed., *Brittle Fracture in Steel Structures,* Butterworths (London), 1970.
2. A.D.S. Carter, *Mechanical Reliability*, Macmillan (New York), 1972.
3. G.E. Dieter, *Mechanical Metallurgy*, McGraw-Hill (New York), 1977.
4. R.A. Flinn and P.K. Trojan, *Engineering Materials and Their Applications*, Houghton Mifflin (Boston), 1975.
5. G. Koves, *Materials for Structural and Mechanical Functions*, Hayden (New York), 1970.
6. *Metals Handbook*, Vol. 1, 'Properties and Selection of Metals', 8th Edn, American Soc. for Metals, 1961.
7. J.B. Moss, *Properties of Engineering Materials*, Butterworths (London), 1971.
8. B.W. Niebel and A.B. Draper, *Product Design and Process Engineering,* McGraw-Hill (New York), 1974.

8

Economic Fundamentals of Selection

8.1 INTRODUCTION

The engineer is usually confronted with two main environments, the technical and the economic. His success in employing the technical environment in order to manufacture products or services depends on his knowledge of technical and physical laws. However, the value of his products and services lies in their utility measured in economic terms. There are numerous examples of materials and structures that have excellent properties, and processes and machines that exhibit excellent performance, but possess little economic merit and consequently have not been adopted industrially. For this reason, it is essential that engineering proposals should be evaluated in terms of worth and cost before they are undertaken. An essential prerequisite of a successful engineering application is economic feasibility.

The important factors that influence the decision whether or not to adopt a new product can be grouped as follows: 1, Product variables, which include the product's utility value, the need for it, its sales appeal, the size of the potential market, and its expected life. 2, Company's variables, which include the suitability of the company's engineering talent and production facilities, the suitability of the company's sales force and means of distribution, and the compatibility of the product with other company products. 3, Cost variables, including research and development costs and setup and tool costs. 4, Competition variables, such as the advantages and improvements over similar products on the market and the strength of the company compared to the competition. 5, The profit potential. The profit factor is crucial in making the final decision in adopting the product. Only through a knowledge of costs can profits be made, and since profits are the reason for the existence of business, a knowledge of costs is most essential.

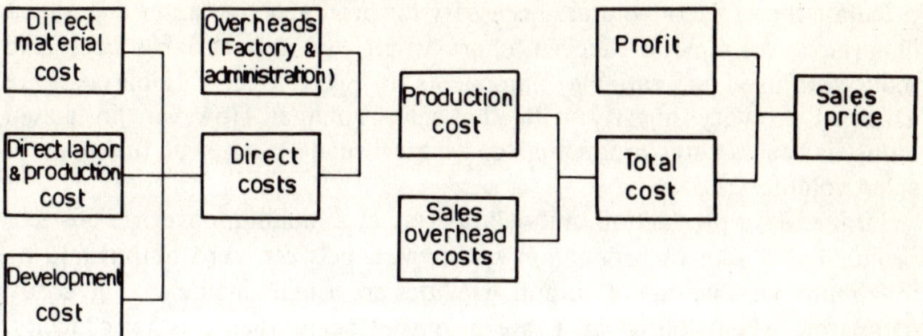

FIG. 8.1. Costs encountered in manufacturing and selling a product, and their relation to sales price and profit.

An understanding of the basis of costs will enable the engineer to use good judgement in the selection of materials and processes to create the optimum product. Figure 8.1 shows the different costs that are usually encountered in manufacturing and selling a product, and their relation to sales price and profit.

8.2 TYPES OF COST

Costs are usually broken down into either two or three broad categories: fixed, variable and, sometimes, semivariable. Fixed or overhead costs are independent of the level of output and cover items such as property taxes, administrative expenses, and building depreciation. Expenses that move in close proportion to changes in production are known as variable or direct costs. Direct labour, direct materials and certain supplies used in the production process are covered under this category.

In many operations costs can be categorised as either fixed or variable. Some costs, however, may not fit well into either of these categories but may assume some of the characteristics of each. Indirect labour, temporary personnel and additional administrative expenses necessary to handle heavy work loads are commonly in this class. Semivariable costs may actually be fixed within certain relatively narrow ranges of production volumes, but generally increase in some relation to production or sales.

8.3 BREAKEVEN ANALYSIS

Breakeven analysis makes use of the fixed and variable nature of costs to calculate the range of volumes necessary for profitable operation. Figure 8.2 illustrates a simple breakeven chart where the semivariable costs are reduced to fixed and variable components. In Fig. 8.2, the variable costs are assumed to vary linearly with the sales volume. However, in actual industrial cases total production costs are usually non-linear functions of sales volume.

Ordinarily a production unit will operate at a minimum average cost per product at a rate of production somewhere between zero output and its maximum possible rate of output. Facilities are usually inefficient and costly to operate when utilised at a rate of output below their normal capacity. Overloading of equipment will result in increased production at the expense of power consumption, maintenance cost and shortened life. Generally, normal loading of facilities corresponds to maximum efficiency of operation. The least-cost load determines the level of operation for maximum economy and profit, as shown in Fig. 8.3. Lack of efficiency when facilities are underloaded, and increased depreciation and maintenance costs when facilities are overloaded contribute to the non-linearity of the total costs. In

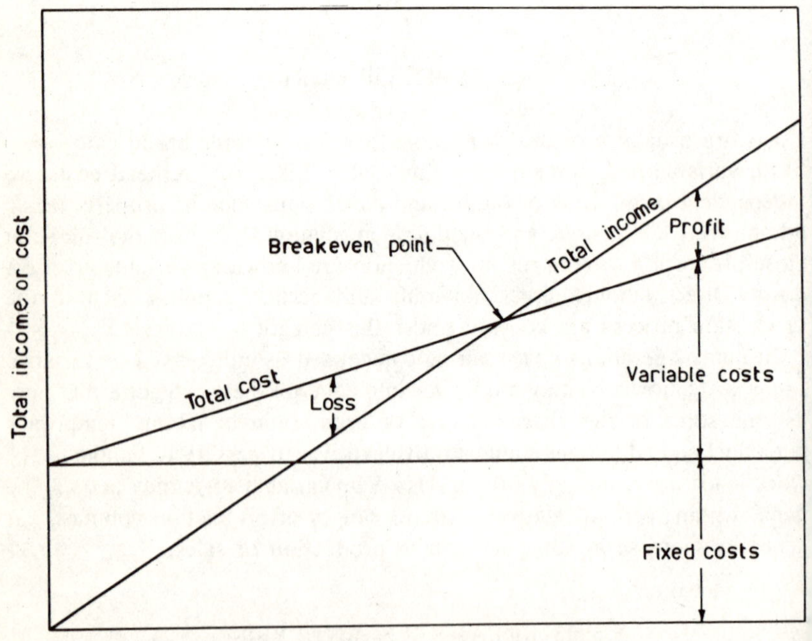

FIG. 8.2. A simple breakeven chart.

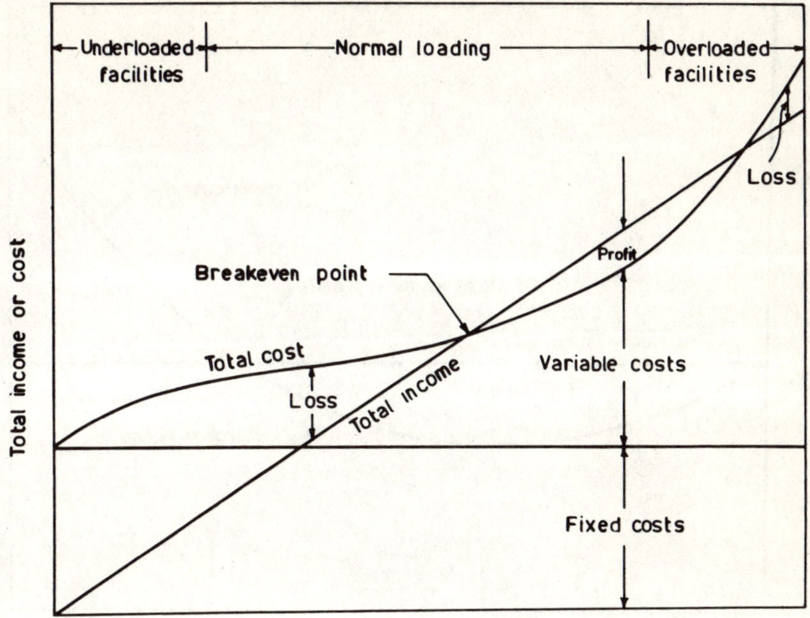

FIG. 8.3. Breakeven chart with non-linear total cost.

these situations it is possible for more than one breakeven point to appear.

The breakeven chart is an important concept for the analysis of many production and process selection problems. It shows the management what will happen to the breakeven point and to the profits as a result of volume changes or proposed courses of action. It also helps in making decisions relating to plant capacity, equipment replacement and make-versus-buy decisions.

8.4 INCREMENTAL COSTS

Incremental, or marginal, costs refer to the additional out-of-pocket costs incurred in producing each additional unit of output without an additional investment in production facilities. In theory, as long as each unit of additional output generates income greater than the out-of-pocket cost of obtaining the added output, it pays to increase production. When, finally,

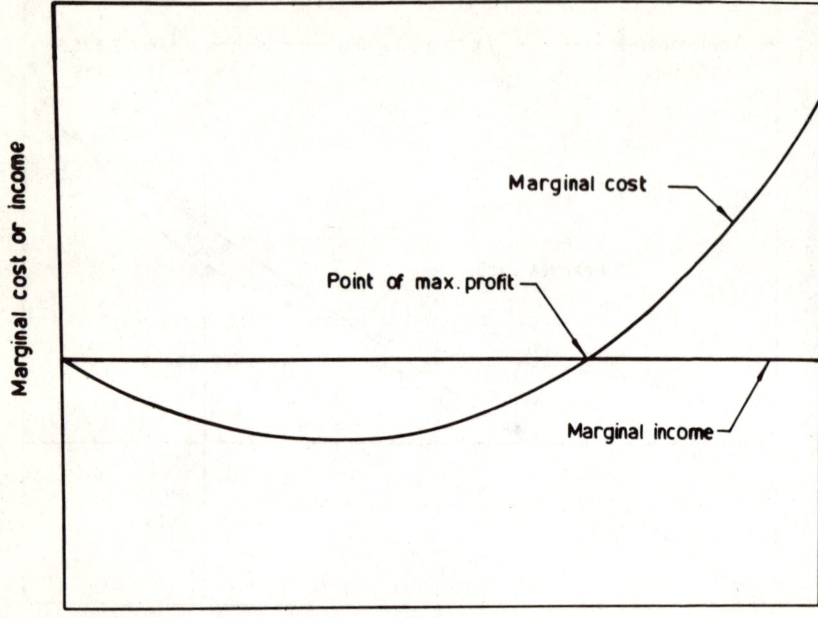

FIG. 8.4. Optimum production level.

marginal cost is equal to marginal income, the optimum level has been reached as illustrated in Fig. 8.4.

If existing production facilities have the capacity to produce a component, the incremental costs will be only the direct costs of labour and materials, plus any actual net additions to other costs such as power and supplies. The machinery, buildings and supervisory and executive staffs already exist; the cost of these does not change in manufacturing the component. Conversely, if a component that is currently being produced is to be stopped, only incremental costs will be eliminated, and these are ordinarily far less than the average manufacturing cost.

8.5 FACTORS AFFECTING PROFIT

Profit, the target for any economic enterprise, is determined by three factors: cost, price and volume. Each of these three factors is in turn influenced by other considerations as shown in Fig. 8.5. The economic environment

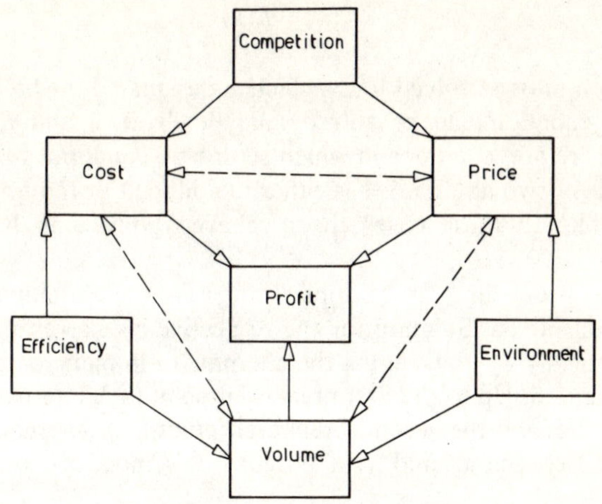

FIG. 8.5. Factors affecting profit.

determines how large the total market is for a specific type of product or service. Competition determines how much of that total market can be obtained at a given price, and the company's efficiency determines how much it costs to produce and market the product or service at the required scale of operation. It is important, therefore, that operations should be planned on the basis of the business climate and outlook, priced on the basis of competitive factors, and carried out as efficiently and as economically as possible.

8.6 THE COST OF ALTERNATIVE SOLUTIONS

In many engineering applications, the cost of a component may be a function of a single variable. When the costs of two or more alternative components are a function of the same variable, it may be desirable to find the value of the variable that will result in equal cost for the alternative components. The cost of each alternative component (T_c) can be expressed as a function of the common independent variable (x) as follows:

$$T_{c1} = f_1(x), \qquad T_{c2} = f_2(x) \tag{8.1}$$

where the suffixes 1 and 2 refer to alternatives 1 and 2 respectively. Solution for the value of x which results in equal cost for alternatives 1 and 2 yields:

$$f_1(x) = f_2(x) \tag{8.2}$$

This equation may be solved for x which is designated the breakeven point. The above equation can be solved analytically if f_1 and f_2 are known. However, there are situations in which setting up equations to represent the cost patterns of two alternatives is either too difficult or too time consuming to be feasible. In these cases the breakeven point may be determined graphically.

The above procedure can be applied to cases where multiple alternatives are being considered. Solution for the respective breakeven points may be found analytically by considering the alternatives in pairs, or graphically as shown in Fig. 8.6. Up to the first breakeven point (x_1), alternative 2 is most economical. Beyond the second breakeven point (x_2), alternative 3 is most economical. Between x_1 and x_2 alternative 1 is most economical.

8.7 MINIMUM COST ANALYSIS

If the cost of a component is a function of a variable that may take on a range of values, it may be useful to determine the value of the variable for which the cost of the component is a minimum. Alternative components whose costs depend upon the same variable can be compared on the basis of their minimum costs. This situation usually arises when the alternatives contain two or more cost factors that are modified differently by the common variable x. Certain cost factors may vary directly with an increase in the value of x, while others may vary inversely.

The general solution for the situation outlined above may be represented as:

$$T_c = \mathrm{a}x + \frac{\mathrm{b}}{x} + \mathrm{c} \tag{8.3}$$

where a, b, and c are constants. The condition for minimum cost can be determined by differentiating Eqn. (8.3) with respect to x, equating to zero, and solving for the value of x:

$$\frac{\mathrm{d}T_c}{\mathrm{d}x} = \mathrm{a} - \frac{\mathrm{b}}{x^2} = 0 \tag{8.4}$$

$$x = \sqrt{\mathrm{b/a}} \tag{8.5}$$

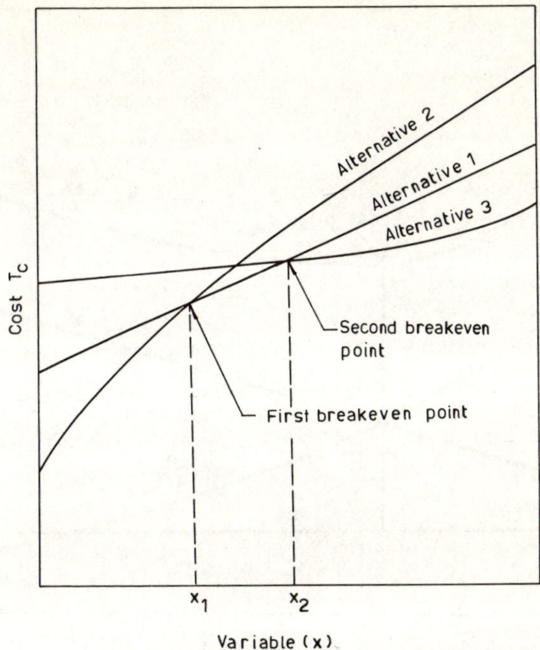

FIG. 8.6. Selection between multiple alternative solutions.

It may be noted that, for the condition of straight line variation of the variable costs assumed, when minimum cost is obtained, the increasing costs are equal to the decreasing cost.

$$\text{Increasing costs} = \text{decreasing costs} = \sqrt{a \cdot b} \tag{8.6}$$

Figure 8.7 graphically illustrates the minimum cost analysis with straight line variation.

A more general solution is where some costs increase with some production or size function, and other costs decrease, but not in direct or inverse proportion as shown in Fig. 8.8. In this general case, the minimum cost point does not have to occur at the point where the two component costs are equal. In many cases the material of a component can be considered as the main independent variable that affects the cost, and a relationship similar to Eqn. (8.4) can be written to relate the component cost (T_c) to its service life (L), weight (W), cost of development effort and pilot runs (R), processing cost (P), assembly and similar costs (A), and material cost (M):

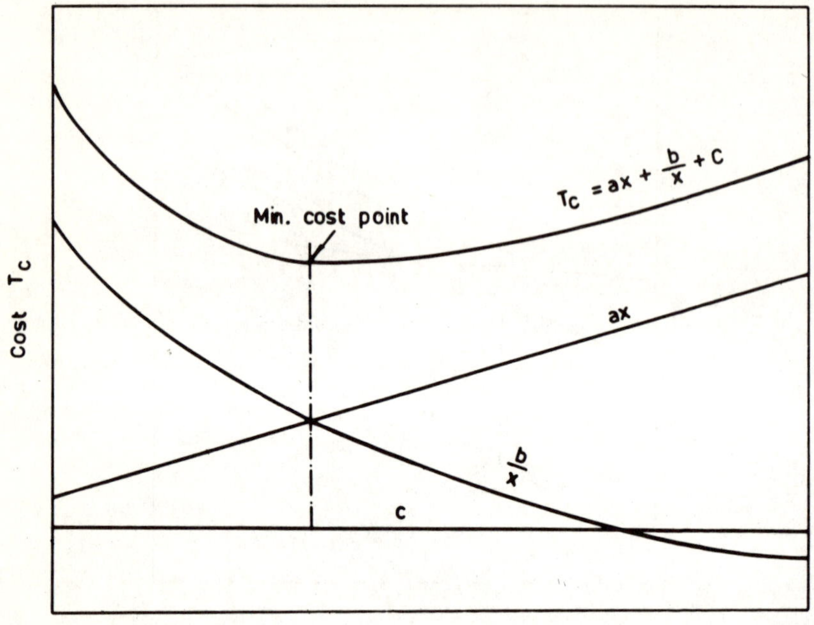

FIG. 8.7. Minimum cost analysis with straight line variation.

$$\frac{dT_c}{dM} = \left(\frac{\partial T_c}{\partial L}\right)\left(\frac{\partial L}{\partial M}\right) + \left(\frac{\partial T_c}{\partial W}\right)\left(\frac{\partial W}{\partial M}\right) + \left(\frac{\partial T_c}{\partial R}\right)\left(\frac{\partial R}{\partial M}\right) + \left(\frac{\partial T_c}{\partial P}\right)\left(\frac{\partial P}{\partial M}\right) + \left(\frac{\partial T_c}{\partial A}\right)\left(\frac{\partial A}{\partial M}\right) + \frac{\partial T_c}{\partial M} \qquad (8.7)$$

When Eqn. (8.7) equals zero the minimum component cost is obtained. In the above equation, the total component cost is assumed to have some functional relationship of the form:

$$T_c = f(L, W, R, P, A, M) \qquad (8.8)$$

It is also assumed that the service life (L), weight (W), research and development costs (R), and processing costs (P) are related to the material costs. This is usually true for many engineering applications. However, Eqn. (8.7) can only be solved if equations relating each variable to material costs are formulated. Usually these relations are known only for large increments.

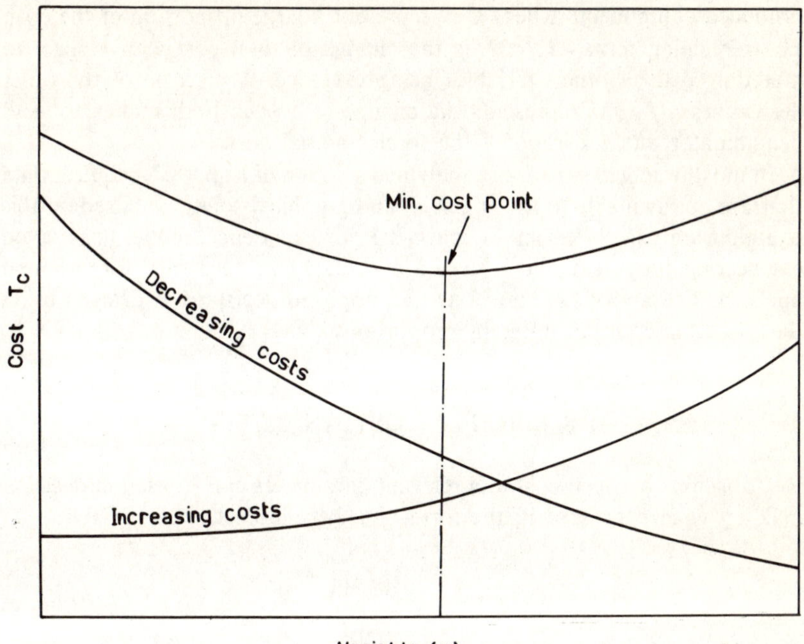

FIG. 8.8. Minimum cost analysis with non-linear cost variation.

Even so, a term-by-term examination of Eqn. (8.7) will be instructive. Depending upon the nature of the component, the analysis is made on a per unit basis or on annual volume basis, whichever is more convenient.

The first term on the right hand side of Eqn. (8.7) is the product of two rates of change, $\partial T_c/\partial L$ and $\partial L/\partial M$. Knowledge of what the consumer will pay for long life or what penalty attaches to short life is necessary to determine $\partial T_c/\partial L$. Data on $\partial L/\partial M$ can be obtained from information on the material's behaviour under expected service conditions. Signs of both parts of this term and the other terms of Eqn. (8.7) are determined by increases or decreases in the variables.

The second term in Eqn. (8.7) may require two or more complete sets of design calculations, in order to establish the relation of total cost to weight $\partial T_c/\partial W$ and the dependency of weight on material cost $\partial W/\partial M$. Cost of the development necessary before the material is adopted for manufacturing the component may be spread over several years, but it must be known in order to find the third terms $\partial T_c/\partial R$ and $\partial R/\partial M$.

The processing and assembly terms are important especially in mass produced components where they represent a large proportion of the cost. The remaining term $\partial T_c/\partial M$, is the change of total cost with respect to material cost, all other variables being excluded. As is true of the other derivatives, $\partial T_c/\partial M$ represents the change of cost in both numerator and denominator, not the ratio of the total material cost.

In most practical situations, analytical solution of Eqn. (8.7) requires data that are not available to the engineer, and graphical solutions based on this equation are usually easier to achieve. Prior experience, model tests, good engineering judgment and accurate cost data are necessary for a sound analysis. The above fundamentals also apply to decisions in designs using the same materials in different proportions.

8.8 BENEFIT–COST ANALYSIS

Relationships between cost and relevant parameters can be used in decision making when expressed in the form of a benefit–cost ratio (BCR):

$$\text{BCR} = \frac{\text{Benefit}}{\text{Cost}} \tag{8.9}$$

BCR can be viewed as an indicator of the price paid for improvements. Benefit in this case covers any improvements in material properties or component performance. When the benefit is expressed in the same units as cost, a BCR = 1 represents the minimum justification for adopting a new material or component in place of the present one.

Considerable care must be exercised in accounting for the benefits and the costs in connection with benefit–cost analysis. Benefits are defined to mean all the advantages, less any disadvantages. Many proposals which embrace valuable benefits also result in inescapable disadvantages. It is the net benefits that are sought. Similarly, costs are defined to mean all costs, less any savings. Such savings are not benefits but are reductions in cost. It is important to realise that adding a number to the numerator does not have the same effect as subtracting the same number from the denominator of the BCR.

When using the benefit–cost analysis to select amongst multiple alternatives, the best solution can be selected by applying the principle of incremental return. First the alternatives are listed in increasing order of cost. The alternative requiring minimum cost, No. 1, is used as the initial base and compared with the next higher cost alternative, No. 2, by computing the incremental benefit (ΔB) and incremental cost (ΔC). If $\Delta B/\Delta C$ is less than unity then alternative No. 1 is better than alternative No. 2 which is rejected, and the comparison proceeds between alternatives No. 1 and No. 3. If $\Delta B/\Delta C$ is greater than unity, No. 1 is rejected and No. 2 becomes the new current best solution and the comparison proceeds between alternatives No. 2 and No. 3. The procedure is repeated until all the alternatives have been rejected in favour of one which is considered to be the most desirable.

8.9 COST-EFFECTIVENESS ANALYSIS

Much of the philosophy and methodology of the cost-effectiveness approach is derived from benefit–cost analysis and, as a result, there are many similarities in the techniques. There are certain steps which constitute a standardised approach to cost-effectiveness evaluations.

First, the desired goals of the system should be defined. Alternative solutions and designs are then developed and a selection is made on the basis of an optimum configuration for each of them. the next step is to establish system evaluation criteria for both the cost and the effectiveness aspects of the system under study. Among the categories of cost are those arising throughout the system life cycle, and these include costs associated with research and development, engineering, test, production, operation and maintenance. System evaluation criteria on the effectiveness side of a cost-effectiveness study are usually difficult to establish. Also many systems have multiple purposes, which complicates the problem further. Some general effectiveness categories are utility, merit, worth, benefit and gain. These are difficult to quantify and, therefore, such criteria as mobility, availability, maintainability, reliability and others are normally used. Although precise quantitative measures are not available for all of these evaluation criteria, they are useful as a basis for describing system effectiveness.

The next step in a cost-effectiveness study is to select the fixed cost or the fixed effectiveness approach. In the fixed cost approach, the basis for selection is the amount of effectiveness obtained at a given cost. The selection criterion in the fixed effectiveness approach is the cost incurred to obtain a given level of effectiveness. When multiple alternatives, which provide the same service, are compared on the basis of cost, the fixed effectiveness approach is being used. This approach is more appropriate when selecting materials or processes for a given application. Candidate systems are then analysed and ranked in the order of their capability to satisfy the most important criterion. Other criteria are ranked in a secondary position. Often this procedure will eliminate the least promising candidates. The remaining candidates can then be subjected to a detailed cost and effectiveness analysis.

Making the best decision possible from the available data can be simplified by adopting the digital logic approach. This requires two steps: 1, the relative importance of each performance requirement is determined; 2, each solution is evaluated against each requirement. For both steps, evaluations are arranged so that only two alternatives are considered at a time. Every possible combination is compared and no shades of choice are required, only a yes or no decision for each evaluation. To determine the relative importance of each performance requirement a table is constructed, the requirements are listed in the left hand column and comparisons are made in the columns to the right as shown in Table 8.1. In comparing two requirements, the more important requirement is given numeral one (1) and the less important is given zero (0). The total number of possible decisions $N = n(n-1)/2$ where n is the number of requirements under consideration. If the total number of positive decisions for each requirement is m, then the summation of the total positive decisions for all requirements (Σm) should be equal to the total number of possible decisions (N). A relative emphasis coefficient (α) for each requirement is obtained by dividing the number of positive decisions for each requirement (m) into the total number of possible decisions (N). In this case $\Sigma\alpha = 1$. This procedure can be modified by giving the more important requirement a number larger than one, depending on its importance, and the less important requirement a number larger than zero if it is relevant to the application under consideration. These modifications are used in Chapters 14 to 20.

To evaluate the different candidate solutions a second table is constructed, and the solutions are evaluated, one against another, for each performance requirement, as shown in Table 8.2. The number of positive decisions for each candidate (P) is then converted to a solution emphasis coefficient (β) for each requirement:

$$\beta = \frac{P}{S}$$

where S is the number of possible decisions. In this case $\Sigma\beta$ for all candidates for each requirement should be equal to unity.

The ability of the different candidates to satisfy the requirements is

TABLE 8.1

DETERMINATION OF THE RELATIVE IMPORTANCE OF PERFORMANCE REQUIREMENTS USING THE DIGITAL LOGIC APPROACH

Require-ments	*Number of possible decisions* $= n(n-1)/2 = 5 \times 4/2 = 10$										*Positive decisions*	*Relative emphasis coeffi-cient* (α)
	1	2	3	4	5	6	7	8	9	10		
Req. 1	1	1	0	1							3	$\alpha_1 = 0{\cdot}3$
Req. 2	0				1	0	1				2	$\alpha_2 = 0{\cdot}2$
Req. 3		0			0			1	0		1	$\alpha_3 = 0{\cdot}1$
Req. 4			1			1		0		0	2	$\alpha_4 = 0{\cdot}2$
Req. 5				0			0		1	1	2	$\alpha_5 = 0{\cdot}2$
	Total number of positive decisions = 10											$\sum \alpha = 1{\cdot}0$

evaluated by constructing a third table where each candidate is given a figure of merit (γ) for each performance requirement:

$$\gamma = \alpha \times \beta$$

where α and β correspond to the candidate under consideration. The most favourable candidate is the one that gets the highest summation of figures of merit for all the performance requirements ($\Sigma\gamma$).

TABLE 8.2

DETERMINATION OF SOLUTION EMPHASIS COEFFICIENT (β)

Candidate solutions	*Requirement 1*								*Requirement 2*	*3*	*4*	*5*
	1	2	3	4	5	6	P*	β_1				
Sol. 1	1	0	1				2	0·33				
Sol. 2	0			0	1		1	0·17				
Sol. 3		1		1		1	3	0·50				
Sol. 4			0		0	0	0	0·0				

*P = Number of positive decisions for each candidate.

In the fixed cost approach to cost-effectiveness analysis the most favourable candidate is the one that gives the highest amount of effectiveness at a given cost. This type of problem is usually met when evaluating public activities involving the general welfare. In this case the granting agency usually has a fixed sum of money and will select the project that will yield the maximum benefit. In the fixed effectiveness approach the most favourable solution is the one that meets the required level of effectiveness at the lowest cost. This type of problem is usually met in industrial applications when selecting materials or processes to perform a given function.

8.10 VALUE ANALYSIS

The objective of value analysis is to determine the most economical way of providing the use values which the consumer requires in a product. Value analysis does not have as its objective the cheapening of products or services by diminishing or eliminating the essential and desirable qualities of proper functioning and safety. This definition of value analysis involves two types of value. The first is 'use value' which is related to the characteristics that accomplish a use, work or service. The second type of value is 'esteem value' which is related to the characteristics that make the consumer want to possess the product or service. These characteristics include appearance and aesthetic acceptibility, reliability, durability and ease of servicing.

A three step approach is often used in value analysis work. The first step involves the identification of primary and secondary functions. The second step is to evaluate the worth of the functions by comparison; and the third step is to develop value alternatives. Grading the value alternatives requires simple economic studies involving selection between alternative designs, materials and manufacturing processes. At this stage esteem values become relevant. An alternative, which might be completely satisfactory from a functional view point, might be lacking in sales appeal. Changes and additions may be necessary in order to provide the required esteem value at a minimum cost.

Depending on the type of industry and on the product, the value analysis assessment will be performance oriented or value oriented. Military and aerospace applications usually involve performance oriented value analysis, at least in the initial designs. Materials and production methods are used almost entirely to assure the required ultimate performance. The value oriented approach has greater application in the consumer industries. In fact, the mass-production industries depend on being able to produce appliances that satisfactorily fulfil a required function at a price that the general public can afford to pay.

In the field of quality control much can be done by applying the principles of value analysis to inspection procedures. In fields where a component failure would endanger human life, 100% inspection is necessary; while in less vital fields, it may be found that better value is obtained by testing the finished assembly and neglecting the inspection of components or sub-assemblies. In the latter case, the cost saved could be more than the cost of the occasional finished appliance that is rejected at the final inspection. Generally, it is useful to review at frequent intervals the cost of an inspection operation, and compare it with the cost that would have been incurred if the operation had not been carried out, and the rejects had been found at a later stage.

BIBLIOGRAPHY

1. E.S. Buffa, *Modern Production Management,* 4th Edn, John Wiley and Sons (New York), 1973.
2. L.W. Crum, *Value Engineering*, Longman (London), 1971.
3. E.P. De Garmo and J.R. Canada, *Engineering Economy*, 5th Edn, Macmillan (New York), 1973.
4. E.D. Heller, *Value Management: Value Engineering and Cost Reduction* Addison Wesley (Reading Mass., London, Amsterdam), 1971.
5. W.R. Park, *Cost Engineering Analysis*, John Wiley and Sons (New York), 1973.
6. K. Seiler III, *Introduction to Systems Cost-Effectiveness*, Wiley-Interscience (New York), 1969.
7. H.G. Thuesen, W.J. Fabrycky and G.J. Thuesen, *Engineering Economy*, 4th Edn, Prentice-Hall (New Jersey), 1971.

9

Economics of Materials

9.1 INTRODUCTION

Materials have a considerable influence on the cost of engineering products. This is because their cost usually represents a high proportion of the total product cost, and also because their processability affects the level and productivity of the capital and labour required for their conversion into the final shape. In a wide range of engineering industries, the direct cost of materials represents 30 to 70% of the value of production. As the cost of materials is so important, efforts should be made to optimise their use, to arrive at an overall reduction of the product cost. Using a cheaper material might not be the answer if it is difficult and expensive to process. For example, if the manufacture involves a large amount of machining, it may be more economical to select a more expensive material with better machinability than to select a cheaper material which is difficult to machine.

The stock dimensions can affect the amount of resulting scrap and, consequently, the final cost of the product. The possibility of recycling the scrap is another material factor that can influence the selection process, especially when using an expensive material.

9.2 CLASSIFICATION OF MATERIALS ON COST BASIS

Generally, engineering materials can be classified into two main categories, depending on their cost. The first category contains the commonly used materials which are manufactured by large scale processes to produce them cheaply and more competitively. The second category contains the special or high performance materials manufactured to meet special needs. The materials in the second category are generally more expensive than the materials in the first category, and are only used in order to meet special requirements that cannot be met by the less expensive, commonly used materials. This division must not be too rigid because a material developed for a particular application may prove to have properties that eventually lead to its wide-spread use.

Figure 9.1 gives an approximate comparison of the relative world consumption by weight of some engineering materials. Timber, concrete and steel are the most widely used materials, but the use of polymers and aluminium is expanding rapidly. Compared on a volume basis, the use of polymers is approaching that of steel, as shown in Fig. 9.2. These consumption values closely correspond to the cost of materials, and Figs. 9.3 and 9.4 compare some metallic and polymeric materials, on the basis of cost per unit weight and cost per unit volume respectively. In these figures the cost of hot rolled carbon steel is taken as unity and all the other costs are given relative to it. All forms of plain carbon steels and low alloy steels appear less expensive than other materials when compared on cost per unit weight basis, but some polymers, in view of their lower densities, are cheaper when compared on cost per unit volume basis. The figures show the

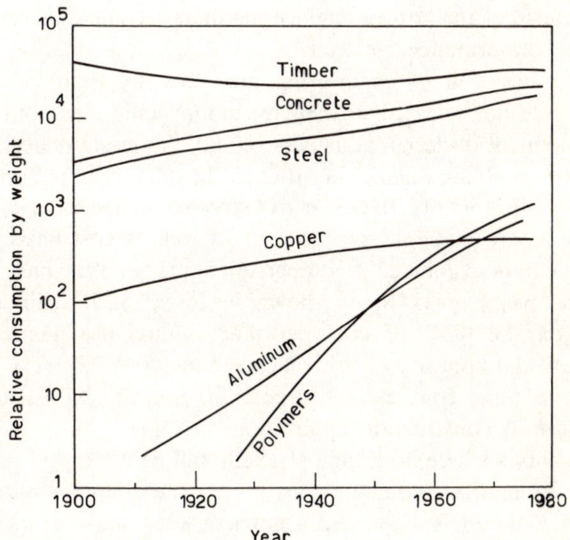

FIG. 9.1. Approximate relative world consumption by weight of some engineering materials.

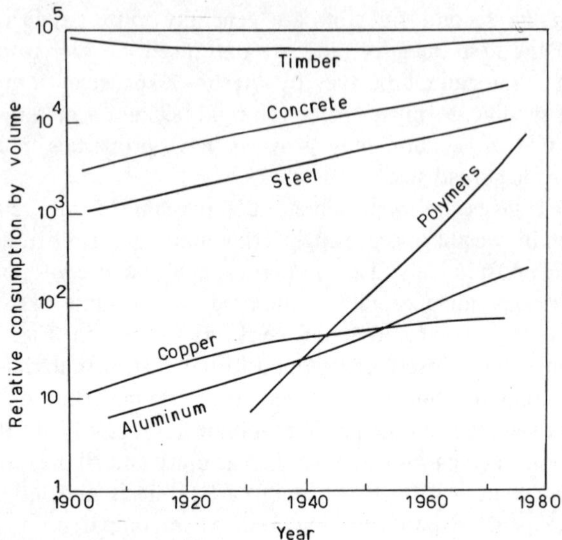

FIG. 9.2. Approximate relative world consumption by volume of some engineering materials.

very wide range of the cost of engineering materials, and illustrate the high cost of high performance materials.

A large proportion of engineering materials are used in applications where they are not stressed to their maximum limits, largely because the prime requirement of design is usually rigidity. So much material has to be used in order to obtain this rigidity that all the other required properties are present in ample amounts. In the case of stressed components, however, it is useful to compare materials on strength as well as cost basis. These two parameters can be combined if comparison is made on the basis of cost per unit volume per unit UTS, as shown in Fig. 9.5. If plain concrete is compared on the basis of cost per unit volume per unit compressive strength, it would appear to be the cheapest material. The very low cost of structural steels and concrete, when compared on this basis, explains their unrivalled use in construction applications.

In applications where both high strength and light weight are important, the specific strength, i.e. strength density, can be used in comparing the cost of engineering materials. Figure 9.6 compares some materials on the basis of cost per unit volume per unit specific tensile strength. On this basis, many polymeric materials, concrete, timber and magnesium castings appear cheaper than structural steels. This comparison explains why polymers, in

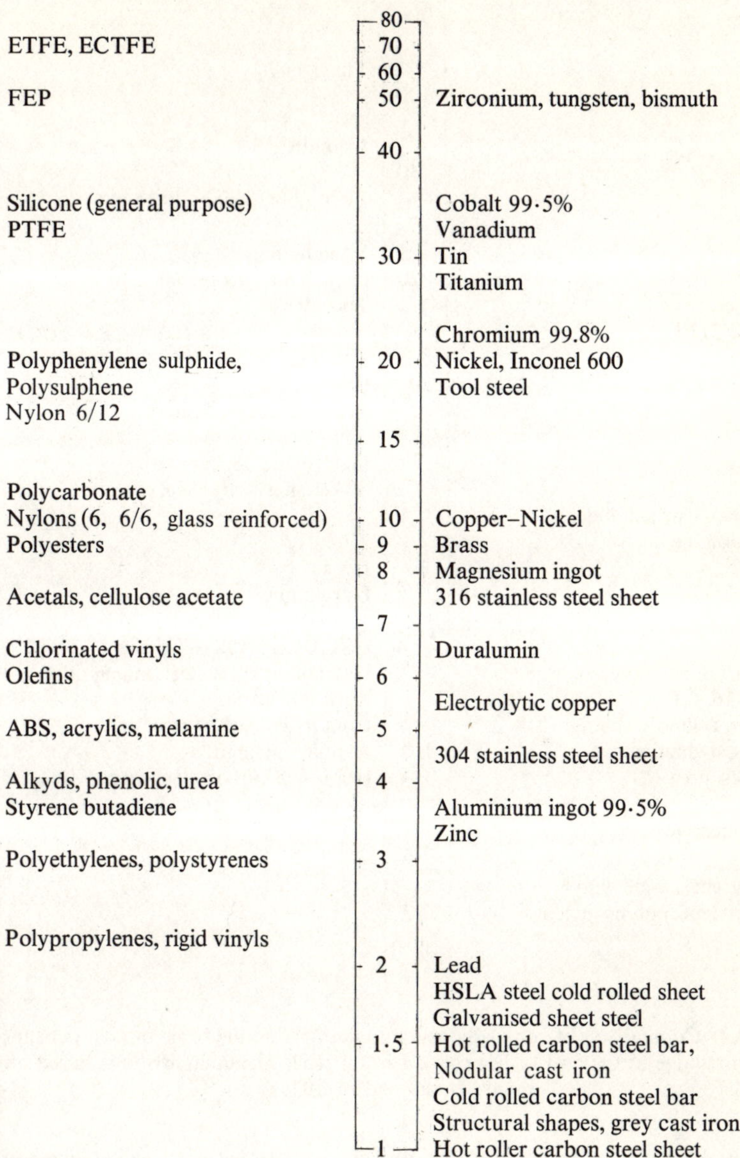

FIG. 9.3. Comparison of some engineering materials on the basis of cost per unit weight relative to the cost of hot rolled carbon steel. The comparison is based on prices at the end of 1976.

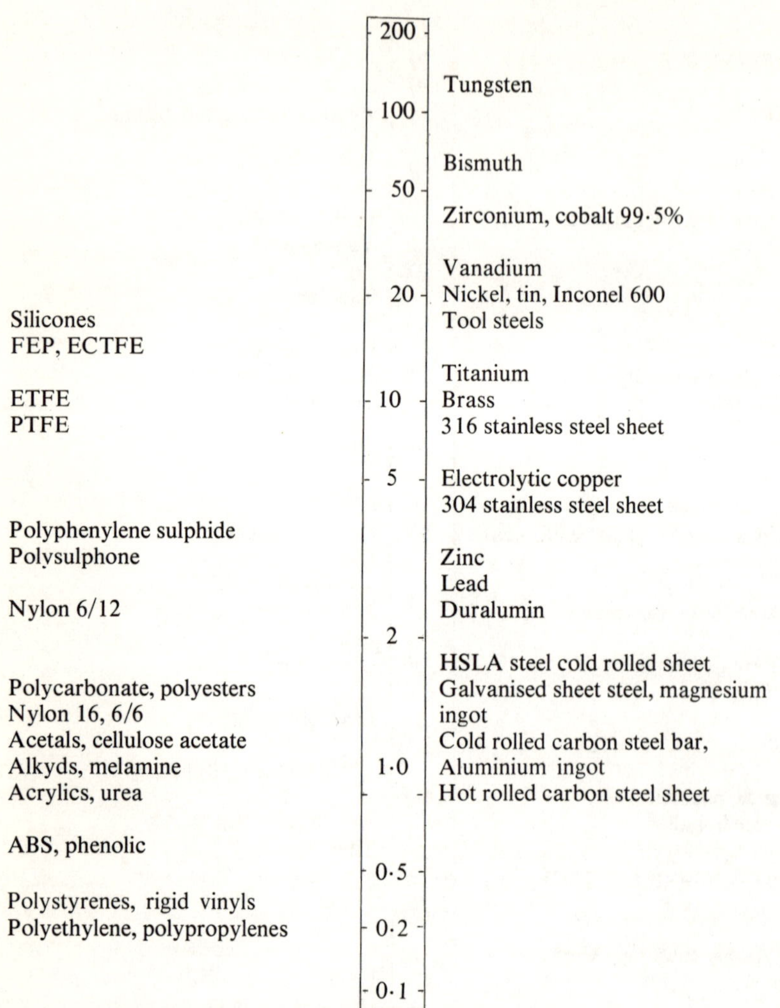

FIG. 9.4. Comparison of some engineering materials on the basis of cost per unit volume relative to the cost of hot rolled carbon steel. The comparison is based on prices at the end of 1976.

spite of their high cost per unit weight, are sometimes used to replace the much cheaper steels. Similar cost comparisons can be carried out to compare materials on the basis of certain required properties, e.g. corrosion resistance, impact strength at low temperatures, high temperature strength or electrical properties.

9.3 ELEMENTS OF THE COST OF MATERIALS

As the cost of materials represents a high proportion of the total cost, it is necessary to analyse its various elements in order to find possible means of minimising it. This can be done by considering the sequence of operations in which raw materials are progressively converted into final products. These

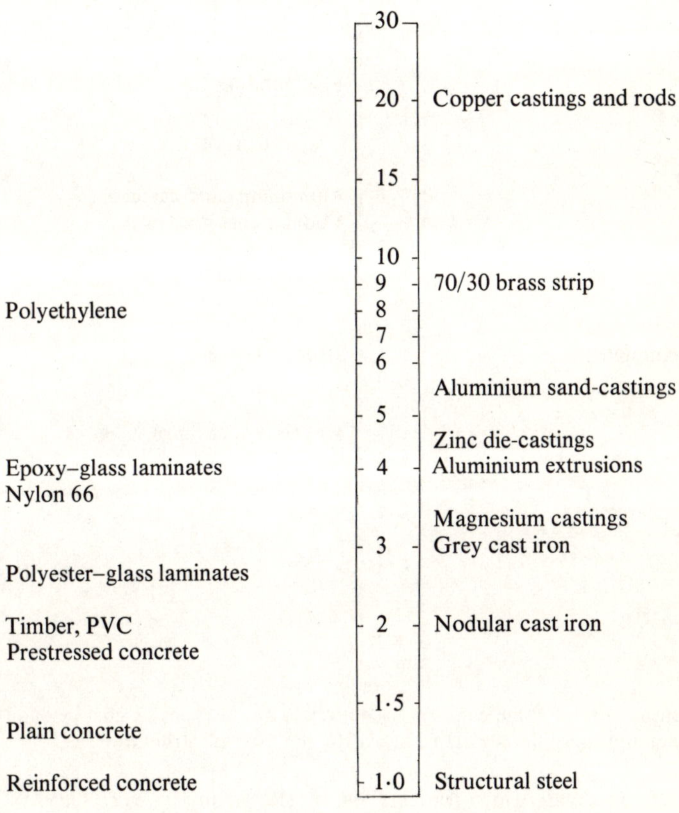

FIG. 9.5. Comparison of some engineering materials on the basis of cost per unit volume per unit UTS relative to the cost of structural steel. The comparison is based on prices at the end of 1976.

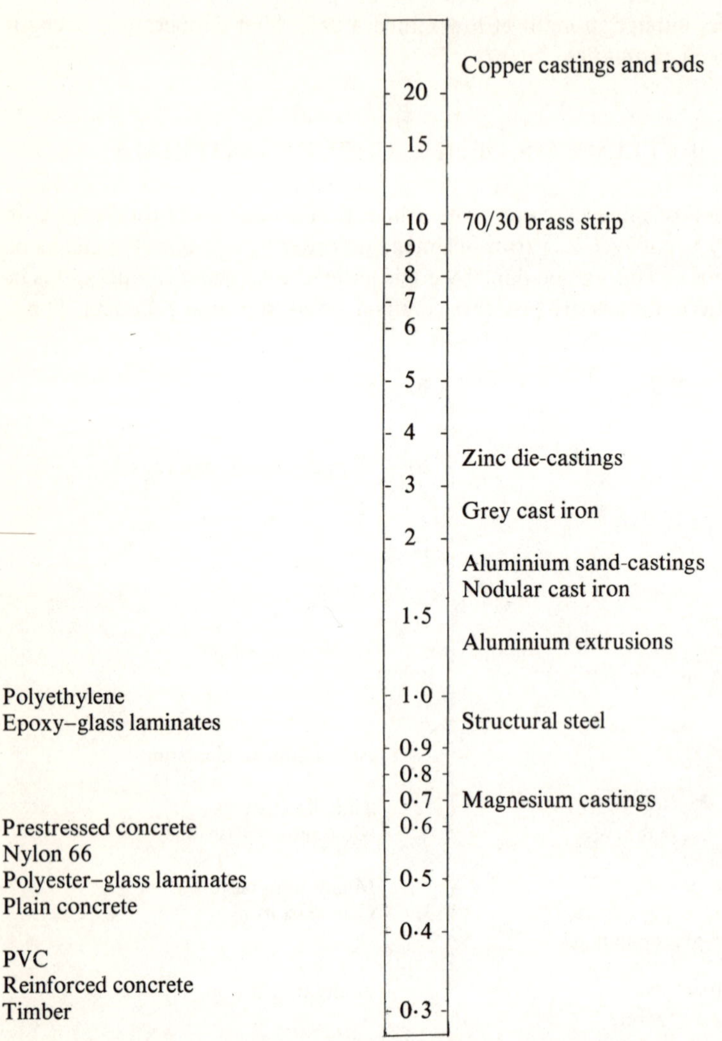

FIG. 9.6. Comparison of some engineering materials on the basis of cost per unit volume per unit specific strength relative to the cost of structural steel.

operations can be divided into four stages, as shown in Fig. 9.7. The first stage is ore preparation for extraction, and the main cost elements are the cost of ore at the mine and the charges for beneficiation. The cost of ore depends on its location and the method of mining it, while the charges for beneficiation depend on the concentration of the required material in the ore and the type of gangue materials.

The second stage is extraction of the material from the ore into ingots, and the main cost elements are the cost of power and cost of auxiliary materials. The large amount of electric power and the high cost of auxiliary materials needed for extraction of aluminium are the main reasons for its high cost (see Fig. 9.7). Another important item of cost at this stage is the addition of alloying elements to the basic material. The cost of alloy is not simply the cost of constituents, because, in the majority of cases, more sophisticated techniques of production have to be employed in order to make full use of the alloying elements. Impurity levels also affect the material costs, and, generally, the higher the permissible impurity level the lower the cost.

The third stage covers the conversion of ingots into semi-finished products ready for delivery to the manufacturer of end products. The main cost elements at this stage are the costs of forming the material in castings, sheets, rods or structural shapes. Another cost element at this stage is due to the material losses. In metal industries these losses range from 25 to 50% in the semi-finishing stages of rolling, forging or casting.

The fourth stage covers the conversion of semi-finished materials into final products by pressing, machining, assembly and finishing processes. These are usually expensive and wasteful operations and material losses can reach 200 to 300%.

In the case of polymers, production begins at the oil or gas well and then it continues in the petrochemical industry for processing the raw materials into monomers, polymers or partially cured thermosetting plastics. Final curing or polymerisation is usually carried out in the same operation that yields the end product. Polymers generally yield a relatively small amount of scrap, which is only usable in secondary products.

The cost of timber in the forest, and the charges for afforestation, correspond to the costs of the first stage in metallic materials production. The felling of timber and dressing of lumber produces waste in large quantities which can be used for by-products such as chipboard and fibreboard. However, the waste material is so cheap that its utilisation is usually a question of transportation costs.

Cement and concrete involve the simplest production cycle with little waste, but concrete has little or no value at the end of service.

At each stage of processing of engineering materials, starting from the

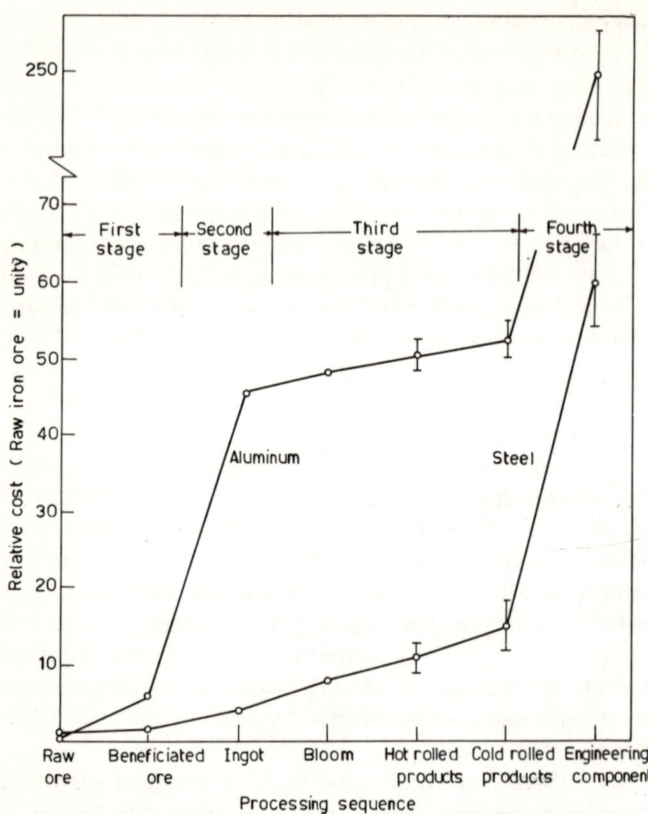

FIG. 9.7. Build up of material cost with the progress of processing operations from ore to finished product.

raw material and proceeding to the finished product, cost is incurred for the energy involved in the conversion or processing, for material losses, for labour and for administrative expenses. All these items of cost go to build up the final cost of the material or product. A major reason for the continuing expansion of the use of polymers, in spite of their relatively high cost, is that by methods such as injection moulding, extrusion, blow moulding and vacuum sheet forming they can be easily converted to usable products without intermediate stages, and with few, if any, finishing processes. These advantages can be gained only with large volume production, and this indicates that the total number of products required may also influence material selection. With mass production, the costs of direct labour and overheads are usually small in proportion to the material costs. Under these conditions it becomes of increasing importance to reduce the material costs and to optimise material utilisation.

Generally, the prices of most established engineering materials show a steady increase according to general economic trends. The main reasons for price increases are rising costs of raw materials, energy and labour, and governmental anti-pollution policies. Over the period 1972–7 most ferrous and non-ferrous metallic materials showed an average rate of price increase of about 7 to 15% per year. Similar rates of price increases are also found for polymeric materials.

9.4 VALUE ASSESSMENT OF MATERIAL PROPERTIES

One of the basic techniques of value analysis is to assess the value of any product by reference to the cheapest available or conceivable product which will perform the same function. This technique can be adapted to material specifications. In the case of steel, for example, plain carbon steels should be considered as a reference point for estimating the value. Additional steel prices above those of plain carbon steel should be critically analysed. The value of each item of a material specification needs to be examined in relation to the function that the finished component has to perform in service.

In comparing materials in order to select the one which will perform the required function at the least expense, the engineer has two basic alternatives: 1, to select the least expensive material, 2, to select a more expensive material that will simplify processing or eliminate steps in manufacturing. An example of cheaper materials that perform the job of more expensive alternatives are in EX steels described in Section 1.4. These steels are designed to replace standard AISI and SAE steels of similar hardenability, but they contain less expensive alloying elements, which makes them more economical to use than the standard steels. Thus in situations where hardenability is the main criterion needed to assure equivalent strengths, an EX steel is usually a suitable replacement. Another example is the manganese-containing grades of stainless steels like 201, 202, 203 and 216, which are less expensive than their type 300 counterparts because they contain less nickel. Cladding, galvanising and tinning are cheaper alternatives to using stainless steels if the corrosion conditions are not severe. Surface hardening or cold working can be used to increase the wear and fatigue resistance of plain carbon steel parts, and the use of expensive alloy steels can thus be avoided.

An example of a higher-cost alternative material resulting in a lower-cost component, because of savings in processing, is prehardened steel sheets. The use of this material makes it possible for users to eliminate heat treatment and the resulting component distortion, as well as the associated handling and salvage operations. Another example is precoated coils which enable fabricators to omit the finishing step. The coating can be alkyd, polyester, acrylic, vinyl, epoxy or phenolic paint, or it can be a plastic, such as vinyl, fluorocarbon or polyethylene. The coatings can be applied to most metals, e.g., steel, tin, zinc and aluminium, and each type of coating provides a different set of properties. Aluminised and chromised coatings on carbon steel resist moderate heat and do not need expensive finishing by enamelling.

9.5 EFFECTIVE USE OF MATERIALS

As materials and processes affect many aspects of production economics, manufacturing a component or a product at a competitive cost can only be accomplished when materials and processes are used as effectively as possible. Effective use of materials will enable the manufacturer to get the highest returns for the money spent. Ideally, the engineer should not pay for properties he does not need and should specify the material that will do the required job at the lowest possible cost, as was discussed in the previous section. Another approach is to develop designs which minimise the amount of expensive materials that have to be used, and to select the processes and stock material forms that maximise the utilisation factor in manufacture.

Selecting the most economical stock can have an important influence on the processing and the cost of a component. Generally materials can be bought more cheaply if ordered in large quantities and in large sizes. Thicker sheets or larger diameter rods are usually less expensive than thinner sheets or smaller diameter rods. Where production involves more than one size, it may become necessary to buy base quantities of large size stock, and use it for a variety of smaller size parts. In some cases the extra machining involved in reducing the large size stock is more costly than buying smaller quantities of the correct size, even though extra costs for small quantity and size must be paid. Calculation of the proper breakeven point with regard to machining costs and quantity and size extras is complicated, and varies with the type of material and the available manufacturing facilities. Surface finish and temper condition of the stock are also among the parameters that can influence the material and processing costs of a component.

Processing of components from powders results in considerable savings because there is little waste of materials, and fewer finishing and machining operations are needed. Using powder paints in coating results in a high utilisation factor which can reach 99%, as opposed to a utilisation factor of 60% for wet spraying of paints.

9.6 COMPETITIONS IN THE MATERIALS FIELD

Earlier discussion has shown that users of materials are constantly searching for alternatives that enable them to maintain the cost at its lowest value without any sacrifice of quality. This creates competition between the primary materials suppliers, and induces them to produce their materials more cheaply and to develop special materials to meet special requirements. The suppliers of major groups of metals are often in competition with each other, and the manufacturers of polymers, ceramics, concrete and timber compete in many fields.

The relative consumption values in Figs. 9.1 and 9.2 show that the use of different materials is increasing at different rates. Generally, the consumption of the older materials like steel, copper, concrete and timber is growing at a slower rate than aluminium and polymers. The slow growth for older materials reflects the competitive inroads made by aluminium and polymers, and also reflects increasing efficiency in their utilisation. The latter case is illustrated by the fact that the use of copper in electrical generators has been reduced from about 100 kg per megawatt to 25 kg per megawatt during the past decade. The more efficient design and the better alloys used in present day aircraft have reduced the use of metals from 3·5 kg per passenger mile in the Boeing stratocruiser to about 1·4 kg in the Boeing 707.

The complex situation resulting from the competition between materials may be illustrated by considering the way in which aluminium and its alloys have advanced in the metallurgical scene. Aluminium, the more recent metal, has been competing with copper in the electrical industry. Overhead cables for power transmission down to 10 kVA, and even lower, are much cheaper when made of aluminium, thus leaving copper to be used mainly for generators and motors because of its higher conductivity per unit cross section and, therefore, its greater compactness in design. The stronger aluminium alloys are usable structurally, having good strength/weight ratios, but considerable changes in design have to be made to allow for their lower elastic modulus. Even the low modulus can be regarded as a virtue in the successful combination of aluminium and steel in making the top deck and life boats of ocean liners and cargo vessels. Apart from the obvious advantage of low density, low modulus helps to keep stresses low and allows the use of medium strength alloys with good weldability and corrosion resistance.

An example of the fierce competition between aluminium and steel is the case of beer cans and oil containers. Aluminium has made an inroad into about one-sixth of this business, but the steel industry has now reduced the thickness of steel in a can to two-thirds in the hope of holding the market. Aluminium is also facing a challenge in the construction of aircraft, and may now yield in critical areas to high strength steels and to the newer titanium alloys and reinforced plastics.

Although plastics are much more expensive than common metals, they are making big inroads in fields normally served by metals. The great attractions of plastics are their low density, the facility with which they can be made into intricate finished shapes, and their attractive colour ranges and aesthetic appeal. These attractions have led to the use of polyethylene and PVC instead of copper, lead and cast iron in water pipes, guttering and tubes. Glass-filled nylon and epoxy resin may be used instead of zinc, magnesium or aluminium die-castings, and even instead of steel and aluminium pressings. Table 9.1 gives a comparison between die-cast zinc and some injection moulded polymers. The zinc die-casting alloy is SAE 903 (0·25% Cu, 3·5–4·3% Al, 0·02–0·05% Mg, balance Zn). Generally, the mechanical properties of the zinc alloy are superior to those of all the plastics under consideration and it is also cheaper on cost per unit weight basis. Zinc is also cheaper when compared on the basis of cost per unit tensile strength and on cost per unit creep strength. This indicates that zinc is less expensive to use in applications requiring strength and dimensional stability. For applications requiring impact and fatigue resistance, however, some polymers become competitive with zinc and even more economical to use, e.g. polypropylene. In applications where light weight is important polymers are more favourable, as shown in Fig. 9.6. Also, in applications where decorative finishes are required, polymers need much less finish and are consequently more economical to use. However, zinc die-castings are more amenable to chrome plating and are favoured for this type of finishing.

TABLE 9.1

COMPARISON OF DIE-CAST ZINC AND INJECTION MOULDED POLYMERS*

Properties	*Die-cast zinc SAE 903*	*ABS*	*Nylon 6/6*	*Polyacetal*	*Polycarbonate*	*Polypropylene*
Tensile strength (MN/m^2)						
at 293 K	260	54	52	63	70	35
at 353 K	220	21	25	33	49	14
Tensile impact strength (relative value)	7·9	1	3·3	1·64	3·6	1·6
Creep resistance expressed as apparent modulus of elasticity at 100 h (MN/m^2)	51 100	1 470	420	1 344	2 030	308
Fatigue flexural strength at 10^7 cycle (MN/m^2)	64·4	15·4	14	24·5	7	18·2
Heat distortion temperature at 1.8 MN/m^2 (K)	$>$ 488	345	324	360	403	332
Change in tensile strength due to 500 h of weathering (%)	+0·5	−8	+10	−34	−11	−30
Specific gravity	6·67	1·05	1·14	1·4	1·2	0·9

(Contd.)

TABLE 9.1 (*Contd.*)

Properties	*Die-cast zinc SAE 903*	*ABS*	*Nylon 6/6*	*Polyacetal*	*Polycarbo-nate*	*Polypropy-lene*
Relative cost/unit weight (zinc = 1)	1	2·1	3·74	4·1	4·2	1·4
Relative cost/unit volume/unit tensile strength at 297 K	1	2·1	3·7	4·5	3·0	1·6
at 353 K	1	3·3	5·3	6·1	3·6	2·6
Relative cost/unit volume/unit creep strength at 297 K	1	11·7	78·3	33·2	19	30
Relative/cost/unit volume/unit tensile impact strength at 297 K	1	2·7	1·5	4·2	1·6	0·9
Relative cost/unit volume/unit fatigue strength at 297 K	1	0·9	1·8	1·7	3·4	0·4

*The values given in this table are derived from Lazar and Herrschaft (see bibliography at end of chapter).

BIBLIOGRAPHY

1. W.O. ALEXANDER, *Contemporary Physics*, **8**, (1967), 5.
2. W.O. ALEXANDER, *Scientific American*, **217** (1967), 255.
3. D.H. BREEN and G.H. WALTER, *Metal Progress*, **102** (1972), 42.
4. A. HOUSTON, K. MISKA and J. MOCK, *Materials Engineering*, January 1976, 69.
5. L.S. LAZAR and D.C. HERRSCHAFT, *Engineering Materials and Design*, April 1972, 303.
6. K.MISKA, *Materials Engineering*, January 1977, 68.
7. H.J. PICK, *Engineering Materials and Design*, November 1972, 3.
8. H.J. PICK, *Engineering Materials and Design*, December 1972, 3.
9. H.J. PICK, *Materials Science and Engineering*, **22** (1976), 189.
10. J.A. VACCARI, *Materials Engineering*, December 1973, 24.
11. C.R. WEYMUELLER, *Metal Progress*, **100** (1971), 71.

10

Economics of Manufacturing Processes

10.1 INTRODUCTION

Discussions in previous chapters have indicated that the properties and behaviour of materials in service are greatly affected by the manufacturing processes used to transform them into finished products. In other words, processing must be regarded as an integral factor in the material selection process. Optimum process selection is an important aspect of component production. Different manufacturing processes have different advantages and limitations; some are ideal for mass production, others for small orders. Some processes require expensive tooling, but produce high precision components requiring little or no further processing, thus reducing overall costs. And some are only applicable to certain materials or product sizes and shapes.

Selecting the optimum process not only avoids production troubles, but directly affects the component cost and marketability. As with materials, the engineer should always be on the look-out for new processes of better efficiency and accuracy or lower cost. Selecting the optimum process is not usually an easy task. Rarely can any product be made by a single process. Several processes are usually available and appear competitive. As a result, optimum process selection can usually be made only after an extensive evaluation and cost analysis of the candidate processes. The evaluation of manufacturing processes should include not only the cost of converting the material into a finished product, but also the material utilisation factor and the possible effect of processing on the material properties and behaviour in service.

10.2 PRODUCTION DESIGN AND PROCESS PLANNING

As distinct from functional design, where the designer's first responsibility is to create a design that functionally meets requirements, production design is aimed at manufacturing a component at the minimum possible cost. This is achieved through the specification of materials, tolerances, basic configurations, method of joining parts, etc. Process planning then attempts to achieve that minimum cost through the selection of processes and their sequence which meet the exacting requirements of the design specifications. This has to be carried out within the limitations of the available equipment, especially in small lot manufacture. If the expected volume of production is large or the design stable, special-purpose equipment may be considered. The sequence of processes involved in the production of a component is usually presented in a flow process chart, in which symbols are used to designate the basic processing steps as shown in Fig. 10.1.

Determining the best sequence of operations is an important step in the processing of a component. Both the cost and quality of the component are closely related to the sequence of operations performed to transform the raw material into a finished product. A different sequence of operations may result in different operation times, different transportation times to the work centre, different tooling and different dimensional accuracy. Operations can be divided into two categories: primary operations and secondary operations. Primary operations include casting, forging, extrusion, etc., where the stock material is formed so as to produce most of the free surfaces of the component. Secondary operations include machining and finishing operations, where the rough form produced by the primary operations is transformed into the finished product.

Generally, the physical location of the facilities of the plant affects the sequence of operations that should be chosen, so as to arrive at the optimum transportation time and distance. The existing loading of the facilities can modify this requirement. Usually locating surfaces and reference surfaces are established early in the sequence of secondary operations, while critical operations involving close dimensional tolerances on fine surface finish are usually done late in the process sequence. Internal operations are usually performed in advance of external operations. Finishing operations are usually carried out late in the sequence of processes, except where prefinished stock is used, e.g. precoated sheets.

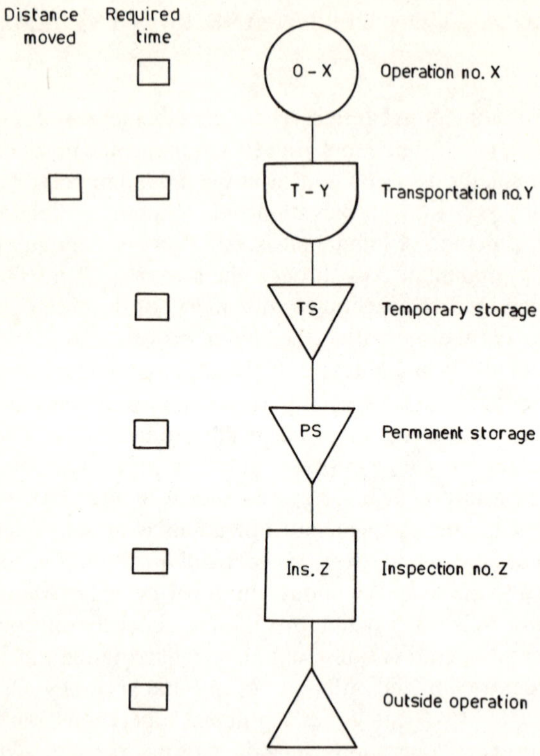

FIG. 10.1. Representation of basic processing steps in a flow process chart.

10.3 MAKE OR BUY DECISION

Every manufacturing concern must decide whether to use its production skill and effort to make each one of a multitude of items or whether to buy them. Even if the decision is to make an item, it still remains to decide on the size, shape, treatment and surface finish of the stock material that will be used. The stock material can be bought in the precoated or pretreated condition to save in processing, or it can be bought more cheaply in the untreated condition and then treated or coated by the manufacturer. The major criterion for decision making in the make or buy area is cost. If a part can be bought cheaper than it can be made, the decision is usually to buy it.

The make or buy decision can be rationalised by the application of minimum cost analysis for multiple alternatives, which was discussed in Chapter 8. The alternative of making may be compared with the alternative of buying if the minimum cost quantity for each is computed and used to find the respective total cost values. Choice of the total cost value that is a minimum identifies the best of the two alternatives. This analysis may be extended to any number of sources; for example, it may be used to compare manufacturing with buying, to compare alternative manufacturing with buying, to compare alternative manufacturing facilities, or to evaluate alternative vendors.

In some cases, factors other than economic ones can influence a given make or busy decision. Quality, reliability and availability of supply; control of trade secrets; patents; flexibility; and alternative supply sources are some of the factors that affect a make or buy decision. It is common to make parts for which basic processes are already available; but when the plant load requires overtime work, the economic analysis may change in favour of buying. A company policy to specialise and to concentrate its efforts and skill in one basic line rather than to diversify may also affect make or buy decisions.

10.4 ECONOMIC ASPECTS OF PROCESS SELECTION

Optimum process selection can be as important as materials selection when designing or redesigning a component. Like materials, manufacturing processes have advantages and disadvantages and it is up to the engineer to select the optimum combination of processes that will yield the least total cost. The word total is an important one since the overall sequence of processes has to be considered, and not just the individual process costs, such as forging costs, machining costs or finishing costs.

Even when a sequence of processes has been selected, the decision remains as to the equipment capacity. An operation can be performed on more than one machine. One may call for a smaller investment, another may do the job faster, and so on. Often the problem is to find which alternative promises the lowest cost to make a given number of components. In that case, the direct costs, applicable indirect costs and fixed costs are computed for each alternative for the number of components required.

An equal cost, or breakeven point analysis, is usually useful for comparing equipment alternatives. Figure 10.2 illustrates the concept of equipment selection based on breakeven analysis. Of course, the breakeven point between any two machines would change for different components and for different materials as the fixed and variable cost elements change. When all the machines under consideration are owned, no capital costs need enter the analysis since these costs exist regardless of which machine is selected. If it were necessary to buy a special machine, for example, the analysis would have to reflect this fact.

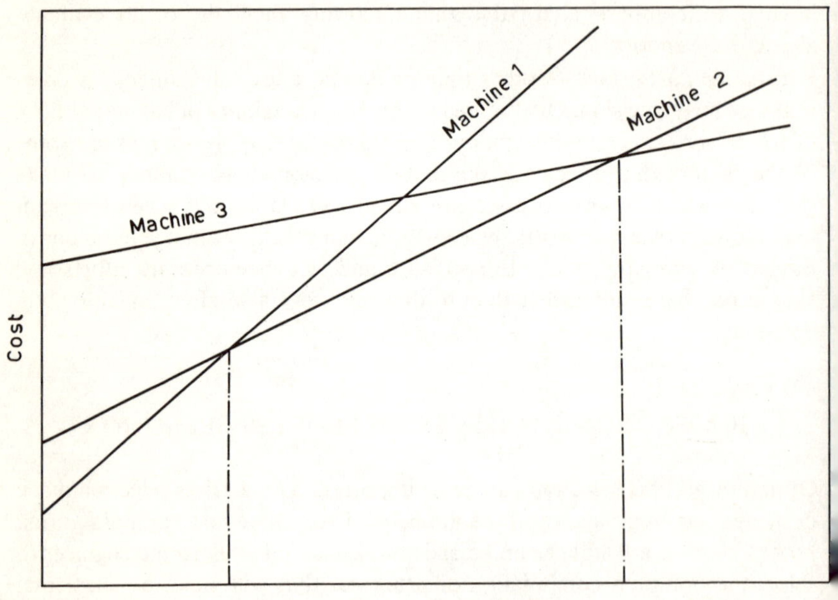

FIG. 10.2. Concept of process selection based upon breakeven analysis. Machine 1 is cheaper to buy but slower and more expensive to operate (e.g. an engine lathe), while machine 3 is expensive to buy but faster and less expensive to operate (e.g. an automatic lathe), and machine 2 is an intermediate case (e.g. a turret lathe).

Although the principle of lowest total cost is a simple one to enunciate, it is more complicated to put into practice because of the difficulty of accurately comparing the relative costs of different combinations of any one material with any particular manufacturing route. This is particularly true in the early stages of designing, and thus it is preferable to identify all possible materials and manufacturing methods before deciding upon the precise details of size, shape, taper, undercuts, etc. of a component. In this way, it is possible to avoid specifying dimensional tolerances that are beyond the commonly accepted limits obtainable by normal manufacturing procedures.

In some cases the selection process may be affected by reasons other than economic ones. The most economical machine to use may be in use or out of order; thus, rather than hold up the order, the next most economical machine can be used. The availability of skilled labour and specific in-house experience account for variations in manufacturing practice from company to company for the fabrication of similar components.

10.5 COST ESTIMATION

The above discussion indicates that estimation of the cost of a product before manufacturing plays an important role in make or buy and process selection decisions. Cost estimation involves determination of variable costs and fixed costs. The variable cost that varies most from machine to machine is labour, which is usually calculated by multiplying the time required for an operation by a labour rate. Thus, the time to set up and perform an operation must be estimated in order to find its variable or direct cost. Other direct costs, such as for power and material, do not usually vary much for the same job when done in different ways. Overhead or fixed costs are commonly calculated by multiplying the operation time by an overhead rate. Such a rate is obtained by dividing the total indirect costs of a production unit for a period of time by the total number of hours of direct labour in the same period.

The total time required to perform an operation may be divided into four parts as follows:

1. *Set-up time.* This is the time required to prepare for operation and may include the time to get tools from the crib and arrange them on the machine. Set-up time is performed once for each lot of parts.
2. *Man or handling time.* This is the time the operator spends loading and unloading the part, manipulating the machine and tools and making measurements during each cycle of the operation. Personal and fatigue allowances as well as time to change tools, etc. are also included in this part. Fatigue allowances depend on the type of work and generally can be considered as proportional to the energy exerted in performing the job. This personal time allows a break from both the physical and psychological stresses that a job may contain and is, in a sense, a minimum fatigue allowance. The minimum allowance is normally 5% of the total available time.

3. *Machine time.* This is the time during each cycle of the operation that the machine is working or the tools are cutting. Many organisations have developed standard data for various machine classes based on accumulated time studies.
4. *Down or lost time.* This is the unavoidable time lost by the operator because of breakdowns, waiting for tools and materials, etc.

The material cost can be based on the weight or quantity of the stock material needed for the part. If applicable, suitable reductions can be made to account for the process scrap.

10.6 PROCESS COST COMPARISON

Generally the choice of a machine for a given job should take into account: 1, the size and shape of the workpiece; 2, the work material; 3, the required accuracy and surface quality; and 4, the quantity of parts and sizes of lots required. Usually a number of available machines can do a job, but the one that will do the job at the lowest cost is the one to be chosen. Of the number of machines that might do a job, most can be eliminated without detailed estimates of their cost, since they are too small or too large, too weak, or in some other ways obviously deficient. The costs of the remaining few can then be ascertained to determine which is the best for the job.

In the case of metal cutting, the power needed at the cutting zone can be estimated by multiplying the rate of metal removal (volume per unit time) by the unit lower in hp/volume/unit time. The unit power varies with the work material, type of operation, cutting angles of the tool, size of cut and speed, but mostly with the first two factors. Accordingly, average values of unit power, like those given in Table 10.1, may be used to estimate the power requirements of common operations. Some loss in power is expected in any machine tool, so the power at the motor must be more than at the cutting zone. An efficiency of about 80% represents average conditions.

Among the other factors that need to be considered in machine tool selection are the accuracy and surface finish which a machine is capable of producing, the removal of hard material or large amounts of stock, the skill available and required to operate the machine, personal likes and dislikes, and availability. As a result of the general considerations just described, all but one size of two or more types of machines can be eliminated from consideration for an operation. It then becomes necessary to estimate and com-

TABLE 10.1

AVERAGE HP PER CM^3 PER MIN OF STOCK MATERIAL REMOVAL FOR SOME MATERIALS

Material	*Hardness (BHN)*	*Process*		
		Turning and shaping	*Drilling*	*Milling*
Steels	85–200	0·17	0·13	0·17
	200–300	0·20	0·16	0·19
	300–375	0·22	0·18	0·20
	375–500	0·27	0·23	0·27
	500–600	0·30	0·27	0·30
Cast irons	110–190	0·13	0·13	0·13
	190–320	0·23	0·22	0·27
Aluminium alloys	30–150	0·03	0·03	0·05
Nickel alloys	80–360	0·30	0·30	0·35
Copper alloys	20–80 RB	0·10	0·08	0·10
	80–100 RB	0·17	0·13	0·17

pare the costs. The general procedure for cost estimation was given in Section 10.5.

Cost comparison between machines can be conveniently carried out using breakeven analysis either graphically, as shown in Fig. 10.2. or analytically as follows. The equal cost quantity N for machines 1 and 2 occurs at:

$$\left(\frac{N}{P_1} + S_1\right)(L_1 + O_1 + D_1) + E_1 = \left(\frac{N}{P_2} + S_2\right)(L_2 + O_2 + D_2) + E_2 \qquad (10.1)$$

or

$$\left(\frac{N}{P_1} + S_1\right) R_1 + E_1 = \left(\frac{N}{P_2} + S_2\right) R_2 + E_2 \qquad (10.2)$$

Where $P =$ number of pieces produced per unit time.
$S =$ time of set-up and tear down.
$L =$ labour cost per unit time.
$O =$ overhead cost per unit time.
$E =$ cost of special tools and equipment chargeable to the job.
$D =$ depreciation per unit time.
$R = L + O + D$

The left-hand-side is the cost of producing N pieces on machine 1, and the right-hand-side is the cost of producing the same number of machine 2. The solution for the equal cost quantity is:

$$N = \frac{P_1P_2\,(S_2R_2 + E_2 - S_1R_1 - E_2)}{P_2R_1 - P_1R_2} \tag{10.3}$$

Although the above discussion is mainly concerned with machine tool and machining process selection, the principles apply equally well to other types of machines and processes if the appropriate data can be estimated.

10.7 TECHNOLOGICAL ASPECTS OF SELECTION

The above discussion shows that process selection is not a simple matter and can only be made when the component function has been analysed in detail. Design alterations can be made in such a way that the form can be changed without affecting the basic function. It may thus be possible to replace an expensive method, using machined parts which have to be assembled, by a cheaper one using a single forming operation with no consequent deterioration in component quality or performance.

Generally, the shape of the component is a useful criterion for eliminating many potential forming routes; e.g. large sheets are generally formed by rolling, tubes formed by drawing, extrusion or centrifugal casting. Typically, investment casting is best suited for complex turbine blades with fine internal channels, and porous materials are only made by powder metallurgy techniques. Fabrication is preferred for low volume, simple structures of large size. Complex, small, or high volume items are usually better made by casting.

In selecting among processes the designer will be interested in the following points: minimum production volumes for a given process; dimensional accuracy and tolerances; the machine finish allowance; as-produced surface finish; component dimension limits and sizes; draft angles; and the possibility of incorporating inserts, forming undercuts and making hollow sections and holes. The alternatives to component forming by casting and forging are machining from a solid blank, and fabrication by joining of preformed sections.

Machining can be judged on the basis of the cost of removing a given volume of metal per unit time and producing a specified surface finish. In view of the relatively high costs of machining operations, the designer should examine the possible introduction of free-machining grades for applications where complexity of design demands large amounts of machining. Machining operations vary significantly in cost; turning and drilling operations are relatively cheap, while milling, gear cutting and grinding are expensive. Since a significant portion of machining costs are due to setting-up times, significant savings are only accomplished when complete machining steps are eliminated. This is particularly true for precision components where parts produced by powder metallurgy techniques compete with castings and forgings in spite of higher powder material costs.

10.8 COMPETITION BETWEEN MANUFACTURING PROCESSES

Discussions in Chapter 9 have outlined the competition in the materials field where, in some cases, more expensive materials have made inroads in fields normally served by cheaper materials. This was shown to be mainly governed by economic reasons, since the use of the more expensive materials can lead to total-cost reductions under certain conditions. A similar competition exists between the manufacturing processes, for example between casting and forging processes, and between casting and fabrication by joining processes.

The two categories of competitive forming methods are casting and forging. In comparing these methods, the basic points to recognise are the influence of die costs and die life on the economic viability of forgings, and the fact that prototype castings can be readily redesigned or modified because of their lower tooling costs. Table 10.2 compares some forging and casting processes on the basis of the minimum quantities needed for a process to be viable when applied to metallic materials. The main factors that control the viability of a process are tooling costs and rates of production; for example open die forging does not require expensive dies but is much slower than closed die forging, which requires expensive dies. Superplastic sheet forming is competitive for relatively small quantities because of low tooling costs, but this advantage is outweighed at high output levels by the long cycle time compared with die-casting and sheet stamping or pressing.

An example of the competition between casting and fabrication processes is the case of machine bodies. When fabrication methods were of inferior accuracy, strength and quality, casting was usually specified. However, improved welding techniques, together with simplicity and elegance of design have allowed cost effective fabrications to compete with castings.

TABLE 10.2

COMPARISON OF SOME FORGING AND CASTING PROCESSES

	Required number of products				
	1–10	*10–100*	*100–1 000*	*1 000–10 000*	*10 000–100 000*
Casting processes					
Sand-casting	0	0	0	0	0
Shell moulding			0	0	0
Pressure die-casting				0	0
Investment casting		0	0	0	0
Forging processes					
Open die forging	0	0	0		
Closed die forging				0	0
Forging of powder preforms				0	0
Superplastic forming	0	0	0		

The tensile behaviour of steel is such that, for a given deflection, a cast iron component would occupy twice the volume. Cast iron, on the other hand, has very much better damping properties than steel. Yet, if a fabricated structure is considered, this advantage is of little value, since the joints contribute the major damping element. Fabrication tends to be more labour intensive than casting and attention to labour costs can make a significant contribution to competitiveness, e.g. by cutting down on welding operations. Since wall thicknesses are greater for cast irons than for fabricated steels, machining operations such as drilling or milling will be cheaper. The advantages of castings, however, are apparent when complex shapes have to be made.

BIBLIOGRAPHY

1. E.S. Buffa, *Modern Production Management*, 4th Edn, John Wiley and Sons (New York), 1973.
2. L.E. Doyle, C.A. Keyser, J.L. Leach, G.F. Schrader and M.B. Singer, *Manufacturing Processes and Materials for Engineers*, Prentice-Hall (New Jersey), 1969.

3. C.K. Singer and M.I. Khan, *Mechanical Costing and Estimation*, Standard Pub. Distributors, Delhi 6, 1969.
4. H.G. Thuesen, W.J. Fabrycky and G.J. Thuesen, *Engineering Economy*, 4th Edn, Prentice-Hall (New Jersey), 1971.

11

Techno-Economic Aspects of Selection

11.1 INTRODUCTION

One of the most important requisites for the development of a satisfactory product at a competitive cost is making a sound economic choice of engineering materials and manufacturing processes. This choice is not a simple task which can be performed at one certain stage in the history of a product, but should gradually evolve during the different stages of product development, as will be discussed in the following sections.

While each material and process selection decision has its own individual character and its own sequence of events, there is a general pattern common to the selection process. This pattern will be described in this chapter and discussed in more detail in Chapter 12.

11.2 STAGES OF PRODUCT DEVELOPMENT

Generally, the development of a new product goes through a series of major phases, that begins with ideas or concepts, and culminates in a finished product or system. The four broad stages of development of a new product are shown in Fig. 11.1, which also outlines an example of the relationship between technical and supporting functions. In the first stage an overall concept is developed and the product functions defined, to find out exactly what the product is supposed to do and what is required of it. In this stage such questions as the following are posed. What is it? What does it do? What are the important design and material requirements? What are the secondary requirements and are they necessary? It is also in the first stage that materials and processing requirements are broadly outlined, and on this

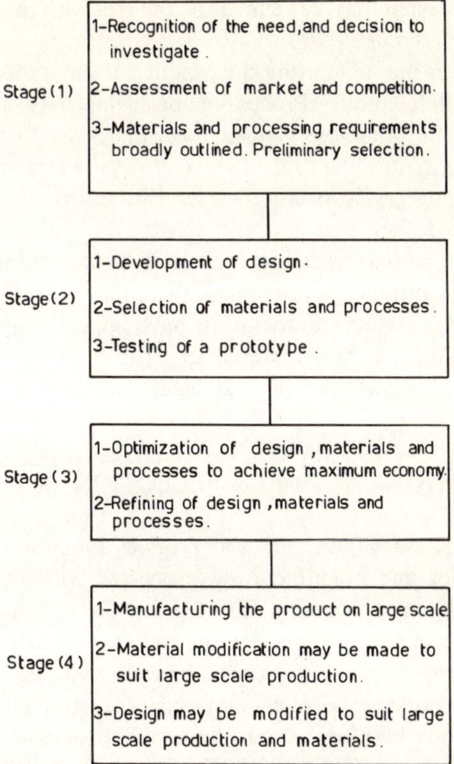

FIG. 11.1. Stages of product development.

basis certain classes of materials may be eliminated and others chosen as likely candidates. Another important part of first stage is the feasibility study where the concept is examined economically in terms of market assessment and competition appraisal.

In the second stage, a practical and workable design is developed and, if necessary, a prototype is constructed. The major part of materials and process selection work is often performed at this stage, and candidate materials are graded according to their expected performance and cost. Processing details are also examined at this stage. In the third stage, optimisation techniques are performed to select the optimum design, material and processing route. At this stage design modifications may be made to achieve production economy or to fit existing facilities and equipment. These modifications could require significant changes in materials, or the design may be refined to conform with the exact properties of the selected materials.

The final stage is manufacturing the product. Even at this stage, materials selection may still continue. Processing problems may arise causing the replacement of an otherwise satisfactory material. For example, heat treating, joining and finishing difficulties may require a material substitution which, in turn, may result in different service performance characteristics requiring some redesign.

Although the materials selection process is most often thought of in the framework of new product development, there may be reasons for material changes in existing products. Some of the most important changes are made to take advantage of a new material or process or to improve service performance, including longer life and high reliability.

11.3 RELATIONS BETWEEN SELECTION PARAMETERS

The above discussion shows that the product function, design, selected material properties and manufacturing processes are closely related. The selection of a design is directly affected by the required function, the material properties and the manufacturing processes, as shown in Fig. 11.2. Secondary relationships also exist between material properties and manufacturing processes, between material properties and the consumer requirements and between manufacturing processes and function requirements. In many cases several alternative designs, materials or manufacturing routes are available, and the basic function of the engineer is to select the most suitable combination. A useful tool in this type of selection problem is cost analysis. Each of the alternative combinations is broken down into individual items which are separately cost analysed and then totalled. In a typical analysis the items are separated into fixed costs and variable costs, as was discussed in Chapter 8. Once the costs are computed then the breakeven point between alternatives is determined. In simple cases it can be calculated simply by the formula:

$$x = \frac{\Delta FC}{\Delta VC}$$

where x is the breakeven point, ΔFC is the fixed cost difference between the two alternatives, and ΔVC is the variable cost difference. Graphical

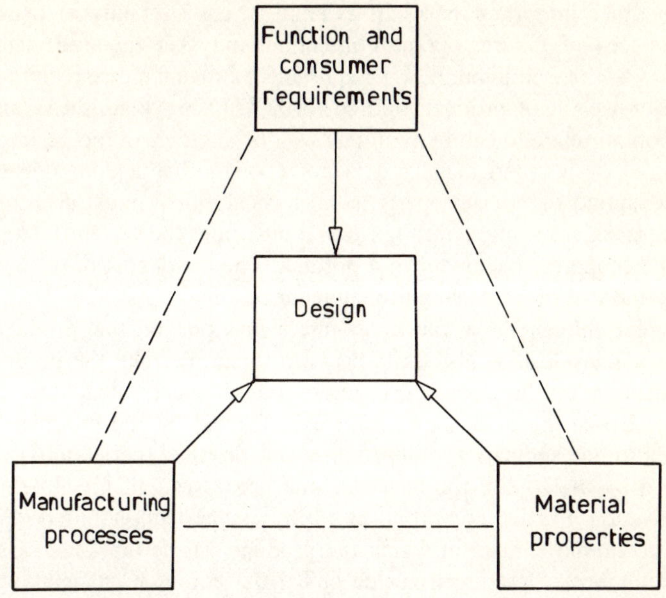

FIG. 11.2 Factors affecting component design.

determinations of the breakeven point can be used in more complicated cases, as discussed in Chapter 8.

In the following sections, the various aspects involved in the selection processes in engineering applications will be discussed in more detail.

11.4 SELECTION AMONG DESIGNS

The technical steps involved in making a new product usually start with design. A successful design should take into account the function, material properties and manufacturing processes, as shown in Fig. 11.2. Besides these factors, a time element should be involved, to take into account the rapid development in materials and manufacturing processes, as well as changing consumer requirements in styling and expected functioning capability. This means that design is a dynamic process in an environment of continuous change.

In a competitive industrial environment, costs are of vital importance and in many cases the cost of a product is a measure of its merit. However, there may be other important bases of evaluation, e.g. certainty of operation, consequences of failure, operator attention and skill required, safety of operator, rate of operation, power requirements, maintenance requirements, volume or weight of product, and so forth. On the whole, these bases of evaluation all relate to economy in one way or another. In fact, economy is inherent in design and usually it is not enough that a component or a machine should function properly as a physical unit; it must also function properly as an economic unit. Figure 11.3 illustrates the various factors that affect the economic behaviour of a product. The initial cost of the product, running and insurance costs and maintenance, repair and breakdown costs have direct influence on the economic behaviour of the product, but secondary relationships also exist between them. To take the product life into consideration, the above cost factors should be presented either as cost per unit service life or cost per service performed by the product.

In consumer industries compromises of physical perfection may be suggested by the cost of the materials and processes which will be needed for production. Further compromises of physical perfection may be dictated by the economic climate in which the product will be marketed. Usually, a large number of products can be built from common elements such as levers, cams, columns, beams, ratchets, gears and pulleys. However, the economic value of each product depends largely upon the appropriateness of the combination of the elements, rather than upon the excellence of the individual elements. This indicates that a holistic approach should be adopted for product design.

11.5 SELECTION AMONG MATERIALS

In recent years, the large increase in the variety of engineering materials, coupled with the rise of new and more severe service requirements and the demand for lower costs, have caused the material selections, for a given component, to be a complex process. The materials selection problem is further complicated by the fact that material properties are influenced by the manufacturing processes and by the geometry of the component and the type of forces acting on it. Frequently, the behaviour of the material in the finished product is quite different from that of the stock material involved in making it. This point is illustrated in Fig. 11.4 which shows the direct influence of stock material properties, production method and component geometry and external forces on the behaviour of materials in the finished

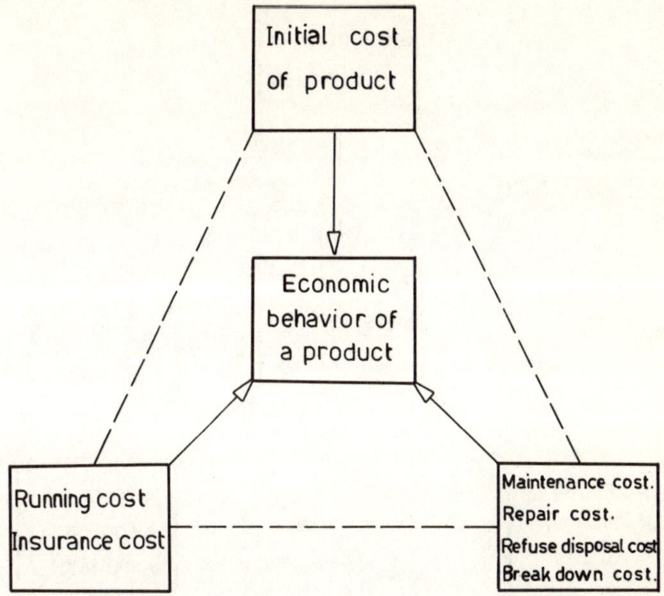

FIG. 11.3. Factors affecting the economic behaviour of a product.

component. The secondary relationships which have to be considered when solving the complex problem of materials selection are also shown in Fig. 11.4.

As component geometry and external forces can directly affect the material behaviour, it is important that the proper design criteria should be used. For example, the presence of sharp corners and fillets can affect the behaviour of high strength steels to a greater extent than mild steels, and surface roughness can affect the material behaviour under fatigue loading to a greater extent than under static loading, as discussed in Chapter 7.

Because materials usually differ with respect to more than one property, true equivalence must be achieved in order to make a judicial comparison between them. The concept of equivalence does not mean that properties must be identical but that the materials may be used so as to produce equivalent results.

From the consumer point of view, it is the performance of the component in service rather than the behaviour of material in the component that counts. Service conditions and component properties in terms of weight, volume, etc. can directly affect the performance, as shown in Fig. 11.5.

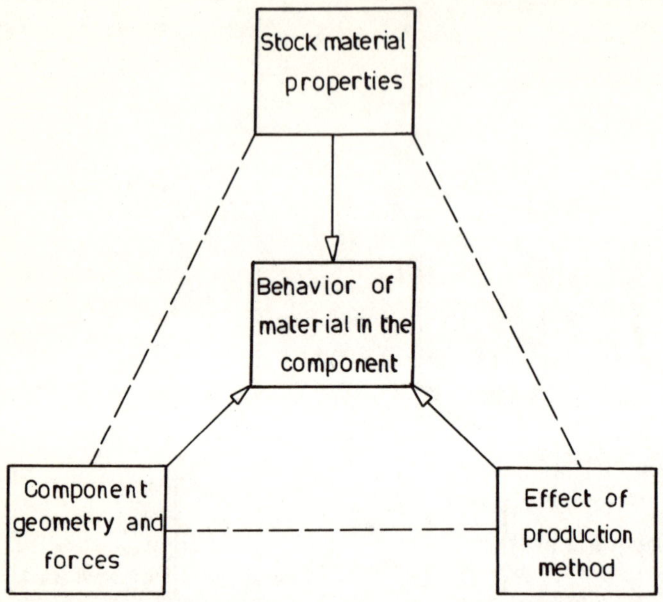

FIG. 11.4. Factors affecting the behaviour of material in the finished component.

In a large proportion of cases, economic selection between materials cannot be based solely on the costs of materials. Very frequently a change in materials will affect the processing costs, and often finishing, inspection and shipping costs may be altered. Care should also be taken to ensure that any differences in yields or resulting scrap are taken into account. Very commonly, alternative materials do not come in the same stock sizes, such as sheet sizes and bar lengths. This may very considerably affect the yield obtained from a given weight of material. Likewise the resulting scrap may differ for different materials. This factor can have serious economic implications when one of the materials is substantially more costly than another.

In determinations of economy, care must be exercised to see that alternative materials provide equivalent services. This, however, can be difficult to achieve especially in comparing qualities. For example, the sales appeal of one type of materials over the other may render their worths non-equivalent. If concrete values can be given to a quality, comparisons can be more easily made. In the case of the quality of lightness for instance, the saving in delivery cost could be taken as a measure in quantitative comparisons between materials. This approach is particularly necessary when applying quantitative methods to materials selection as will be discussed in Chapter 13.

11.6 SELECTION AMONG FABRICATION METHODS

In the cases where several designs and materials may serve a purpose equally well from a functional and cost standpoint, the cost of processing them may be the factor that determines which is chosen. Some designs contain features that inherently are more costly to produce than others. Unnecessarily tight dimensional tolerances are prime causes in this respect. Not only does increased accuracy cost money, but the relaxation of accuracy requirements may make it possible to use different and less expensive processes.

Closely related to the matter of tolerances is the cost of quality. Inspection and control procedures, which are necessary to ensure quality, and the cost of better quality itself, add to the cost of a product. In many

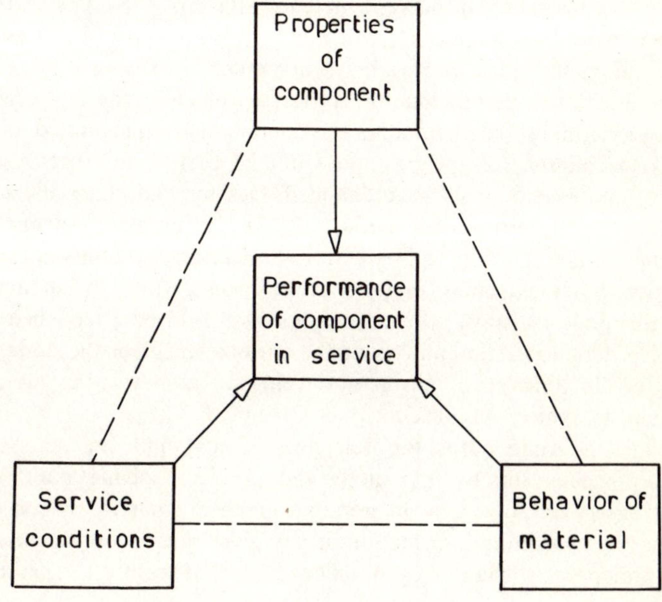

FIG. 11.5. Factors affecting the performance of a component in service.

cases it is more economical to permit a small number of defects to pass the inspection barriers and pay the necessary costs for replacements and repairs on these defective products, than to pay the added cost that would be required to eliminate all defects. Obviously, customer satisfaction is an important factor that must be considered in such decisions.

From the materials standpoint, the cost of processing can play an important role in selection. For example, brass is often found to be less costly for some parts than cold rolled steel because it can be machined at a higher rate, in spite of its greater density and cost. Aluminium, which is easily machinable and in addition has a low density, is being used in increasing amounts as a replacement for steel, cast iron and other metals whose cost per unit weight is considerably less. Because of the ease with which they can be processed, plastics have proved to be economical in many applications as a replacement for materials of less cost per unit weight. This is also the case for die-castings.

In some cases the decision to substitute one material for another will result in an entirely different sequence of processing. For example, a change from grey cast iron to zinc alloy castings will require a marked change in equipment. To determine the comparative economic desirability of two materials it is necessary to make a detailed study of the costs that arise when each is used. In some cases, this may involve the cost of disposing of present equipment and acquiring new equipment.

It was mentioned earlier that the processes of designing and selection should be dynamic to accommodate the continuously changing conditions. The correct solution for the past may not be correct for the present if conditions and requirements have changed. Designs and materials, as well as manufacturing processes, should always be checked against new approaches, materials and techniques. In order for a product to remain competitive, it should continue to give the highest value for money. To achieve this goal, the most efficient conditions should be established at all stages of production. The analysis of the various stages of the design and manufacture in order to obtain greater value for money is the subject of value analysis, which was discussed in Chapter 8.

Although in some cases the selection of materials and fabrication methods proceeds step-by-step along the problem solving path, many practical cases are solved in an iterative manner. That is, in collecting, analysing and evaluating information in any given stage, new insights may be gained or new problems may be uncovered, that require a repetition of earlier steps.

The following chapters outline the techniques and quantitative methods which deal with the various interacting parameters involved in the selection processes in engineering.

BIBLIOGRAPHY

1. E.S. Buffa, *Modern Production Management*, 4th Edn, John Wiley and Sons (New York), 1973.
2. H.R. Clauser, R.J. Fabian and J.A. Mock, *Materials in Design Engineering*, July 1965, 109.

12

The Material Selection Process

12.1 INTRODUCTION

Earlier discussions in this book have shown that one of the most important requisites for the development and manufacture of competitive products is the selection of the optimum combination of materials and processes for the different components of these products. Most modern products are relatively complex and often a variety of materials and processes have to be utilised in order to achieve a proper balance between functional fulfilment, pleasing appearance and reasonable cost. Materials selection is a continuing process and, as described in Chapter 11, starts in the early stages of design and goes on until the final stages of mass production. Even after the component is in production, the process of selection continues in order to make use of new materials and processes. These present possibilities for cost reduction and improved performance. However, new materials must be evaluated very carefully to make sure that all their characteristics are well established. A large number of product failures have resulted from new materials being substituted before their long-term properties were fully known.

With the large number of materials and processes available to the engineer, the selection process is often a difficult task. The problem is further complicated because not only must the material provide the required properties, and not only must the process produce the required shape, but the material and process must be compatible. Also, the fact that the shaping process contributes to the final material properties should not be overlooked.

The general procedure for materials selection can be summarised in the following steps. 1, Analysis of the material requirements. 2, Development of alternative solutions to the problem of materials selection. 3, Evaluation of different solutions and candidate materials. 4, Decision on the optimum material. The different steps will be discussed in the following sections.

12.2 ANALYSIS OF THE MATERIAL REQUIREMENTS

The outcome of the design stage in product development is a detailed specification of the material performance requirements, which are based on the desired function and the expected performance of the component, as well as the environment in which it will operate. These material requirements are then translated into mechanical, physical and chemical material properties. In some cases, this is relatively easy, as for example in components where uniaxial stresses are involved. In this case, the yield strength of a candidate material can be compared with the stresses encountered in the component. However, some specifications may not have simple correspondence with measurable material properties as in the case of wear resistance, weldability or reliability, for instance. Under these conditions, the evaluation process can be quite complex and may depend upon predictions based on simulated service tests or upon property tests most closely related to the service requirements.

The material performance requirements of any product can be divided into five broad categories:

1. Functional requirements,
2. Processability requirements,
3. Cost,
4. Reliability,
5. Resistance to service conditions.

12.3 FUNCTIONAL REQUIREMENTS

The functional requirements are directly related to the required characteristics of the product. It is not always possible to assign quantitative values to these product characteristics. Nevertheless, in one way or another, they must be related as precisely as possible to the most closely applicable mechanical, physical or chemical properties.

12.4 PROCESSABILITY REQUIREMENTS

The processability of a material is a measure of its ability to be worked and shaped into a finished component. With reference to a specific manufacturing method, processability can be defined as castability, weldability, machinability, etc. Ductility and hardenability can be relevant to processability if the material is to be deformed or hardened by heat treatment respectively. The closeness of the stock form to the required product form can also be taken as a measure of processability in some cases.

It is important to remember that processing operations will almost always affect the material properties so that processability considerations are closely related to functional requirements.

12.5 MATERIALS COST

Cost is usually the controlling factor in evaluating materials because in many applications there is a cost limit for a material intended to meet the application requirements. When the cost limit is exceeded, the design is changed to alter the material requirements to keep below the cost limit.

Although materials are usually priced in terms of cost per unit weight, comparisons are better made on the basis of cost per unit volume or cost per unit relevant property as was discussed in Chapter 9.

The cost of processing often exceeds the stock material cost and in many cases a relatively more expensive material may eventually cost less than a low-priced material that is more costly to process.

12.6 RELIABILITY REQUIREMENTS

Reliability of a material can be defined as the probability that it will perform the intended service for the expected life without failure. Material reliability is difficult to measure, because it is not only dependent upon the material's inherent nature and properties, but is also largely a function of its production and processing history. Generally, new and non-standard materials will tend to have lower reliability than established, standard materials.

Despite the difficulties of evaluating reliability, it is often an important selection factor that must be taken into account. Failure analysis techniques are usually used to predict and anticipate the different ways in which a product can fail, and can be considered as a systematic approach to reliability evaluation. The causes of failure of a component in service can usually be traced back to defects in materials and processing or to unexpected service conditions. The subject of failure analysis has been dealt with in some detail in Chapter 7.

12.7 SERVICE REQUIREMENTS

The environment in which the component will operate plays an important role in determining the material requirements. Corrosive environments as well as high or low temperatures can alter adversely the service performance of most materials, as was discussed in Chapter 6. Whenever more than one material is involved in an application, compatibility becomes a selection consideration. In thermal environments, for example, the coefficients of thermal expansion of all the materials involved in a part may have to be similar to avoid thermal stresses. In wet environments, materials that will be in physical contact should be chosen carefully to avoid galvanic corrosion.

In applications where relative movement exists between different components, wear resistance of the materials involved should be considered. The design should provide for access for lubrication, otherwise self-lubricating materials have to be selected.

12.8 CLASSIFICATION OF MATERIAL REQUIREMENTS

After the above analysis of the material requirements, the next step is to classify these requirements into two main groups. These groups are: 1, 'Rigid' or go–no–go requirements, and 2, 'Soft' or relative requirements. Rigid requirements are those which must be met by the material if it is to be considered at all. Examples of rigid requirements are availability and processability which are parameters that do not allow compromise. Thus, if a material is not readily available, or cannot be formed into the desired shape, then it should not be considered in the first place. Rigid requirements can also include those which specify minimum performance characteristics. Any merit in excess of this minimum level is of no special advantage, nor would it make up for any oher missing qualities. Compatibility of the material with neighbouring materials is an example.

Soft or relative requirements are those which are subject to compromise and trade-offs. Examples of soft requirements are mechanical properties, specific gravity and cost. Soft requirements can be compared in terms of their relative importance, which depends on the application under study.

Some material requirements can be classified either as rigid or soft requirements depending on the application for which the material is intended. For example, in the case of gas turbine blades a high melting point is a rigid requirement but it is not in the case of an electrical insulator. On the other hand, high electrical resistivity is a rigid requirement for the insulator but not for the turbine blades. Examples of rigid and soft requirements will be discussed in detail for a wide range of engineering applications in Chapters 14 to 20.

12.9 DEVELOPMENT OF ALTERNATIVE SOLUTIONS

Once the material requirements are clearly specified and classified, the rest of the selection process involves the search for the material or materials that best meet those requirements. The starting point for materials selection is the entire range of engineering materials.

In this phase of the selection process different approaches to the solution of the materials problems, as distinct from simply choosing candidate materials for evaluation, should be considered. At this stage creativity is essential in order to open up channels in different directions and not to let traditional thinking interfere with the exploration of ideas. The importance of this phase is that it creates alternatives without much regard to their feasibility. After all of the alternatives have been suggested, the ideas that are obviously unsuitable are eliminated and attention is concentrated on those that look practical. Past experience and available production facilities play an important role at this stage. At the end of this phase, the number of candidate materials is narrowed down to a manageable number for subsequent detailed evaluation. The classification of the material requirements into rigid and soft requirements can also simplify the initial screening process. Comparisons of the different properties of the metallic, polymeric, ceramic and composite materials, given in Chapters 1 to 7, would be of help in illustrating the differences between engineering material classes. Cost comparisons are given in Chapter 8.

12.10 EVALUATION OF ALTERNATIVE MATERIALS

Having narrowed down the field of possible materials to those that do not violate any of the rigid material requirements, the search starts for the material, or materials, that best meet the soft material requirements. The objective of the evaluation stage is to weigh the candiate materials against the specified material requirements in order to select the optimum one for the application. Unlike the sciences where there is normally only a single correct answer to each and every problem, materials selection requires the consideration of conflicting advantages and limitations, necessitating compromises and decision making; and as a consequence different solutions are possible. This is illustrated by the fact that similar components performing similar functions, but produced by different manufacturers, are often made from different materials and even by different manufacturing processes.

In practice there is no one formal procedure for materials selection. The most appropriate approach will, among other things, depend on the nature of the application, the company, and the personal preference of the engineer. The evaluation phase may begin with the most critical property and then proceed to those of lesser importance; or the candidates may be compared on the basis of all the pertinent properties. In many applications, rating systems and quantitative methods can be used to advantage. Chapter 13 gives a detailed account of several quantitative methods of materials selection. Whatever the approach used for evaluation, it has to take into account the effect of manufacturing processes on material properties and on the material, processing and finishing costs.

12.11 DECIDING ON THE OPTIMUM MATERIAL

The final step in materials selection involves the decision on the optimum material. In many cases the evaluation techniques that will be discussed in detail in Chapter 13 indicate the best material to use and the decision is easy. However, when the results of evaluation are not clear cut, the engineer should use his judgement and experience in making the final decision. In this case, the evaluation techniques can be used in an iterative mode by changing the relative importance of the different soft material requirements. When none of the available materials are found to meet the requirements, either the specifications should be relaxed or other designs must be considered.

12.12 AIDS TO MATERIALS SELECTION

Engineers involved in materials selection are now confronted with such a large amount of materials information that only the broad outlines of the selection process can be given here, and the engineer will need to have ready access to other sources of data. These are numerous, and Table 12.1 lists some of the widely available periodicals and references. An article in *Materials Engineering*, **11**(1975) 20–32, lists many addresses for materials information sources in the USA. Government agencies, trade associations,

TABLE 12.1

MATERIALS INFORMATION SOURCES

Information source	*Publisher*	*Characteristics*
1. *Materials Selector*	Issue of *Materials Engineering,* USA.	One issue each year of the monthly periodical. Contains tabulated data and advertising about common engineering materials.
2. *Data Book*	Issue of *Metal Progress,* ASM, USA.	One issue each year of the monthly periodical. Contains tabulated data regarding materials, process engineering and fabrication technology.
3. Volume 1 of the *Metals Handbook*	ASM, USA.	This volume deals solely with the properties and selection of metallic materials.
4. *ASTM Standards*	ASTM, USA.	An authoritative source for information on materials, properties and test techniques.
5. ASTM Information Retrieval Service	ASTM, USA.	Available free of charge from ASTM information centre.
6. ASM Computerised Data Base	ASM, USA	Comprehensive system for access to metals literature from 1966 to the present.
7. Alloy Index in *Metals Abstracts*	The Metals Society, UK.	Lists all references to specific alloys mentioned in *Metals Abstracts.*

engineering societies, research institutes and materials producers are listed in that article.

One of the biggest hurdles in conducting a large scale evaluation of materials and processes is the sheer amount of data that has to be processed. This data has to be assembled, analysed and classified into a useful form. Many computer aided systems are now available to simplify this task. The heart of a computer aided materials selection system is a computerised data bank with the capability of storing data on different engineering materials. This stored data must be kept up to date if it is to be useful. The data bank should also include information on costs and processing.

12.13 ORGANISING FOR MATERIALS SELECTION

Earlier discussions have shown that the procedure of materials and process selection is complex and, because it runs through the entire design, development, production and operational history of a product's life, different personnel and departments become involved. This can create problems of liaison and control. The most common method of providing the necessary continuity and control throughout all the product development phases is to organise on a project basis and to set up a committee or team for each project. The size, composition and nature of the project group depend on the product involved. If the product is simple and the company is small, one project engineer may perform all of the many required functions. More often, however, a number of individuals will form the project team, and one or several will be responsible for handling the materials selection problems. Other members of the team are representatives from the various departments or areas concerned with the product design, development and manufacture.

In the project team, design engineers establish performance specifications and reduce technical findings to a practical design. Materials engineers examine material performance requirements against available materials. The manufacturing engineers anticipate possible manufacturing problems and conduct the necessary process development work.

BIBLIOGRAPHY

1. H.R. Clauser, R.J. Fabian and J.A. Mock, *Material in Design Engineering*, July 1965, 109.
2. E.P. De Garmo, *Materials and Processes in Manufacturing*, 4th Edn, Macmillan (New York), 1974.
3. P.M. Unterweiser, *Metal Progress*, **111** (1977), 38.

13

Quantitative Methods of Materials Selection

13.1 INTRODUCTION

Discussions in Chapter 12 have shown that the materials selection process can be a very complex operation in which a large number of materials, often covering a wide range of properties, have to be compared in order to arrive at the optimum material. If the selection process is carried out haphazardly, there will be the risk of overlooking a possible candidate material. This risk can be reduced by adopting a systematic materials selection procedure. The quantitative procedures which will be presented in this chapter have been developed to analyse the large amount of data involved in materials selection so that a systematic evaluation can be made. All the procedures presented here can easily be adopted for computer aided selection from a data bank or an information retrieval device.

It should be emphasised at this stage that none of the given quantitative procedures is meant to replace the judgement and experience of the engineer. The procedures are only meant to help him in making sounder choices and trade-offs when selecting materials and processes for a given application.

13.2 COST PER UNIT PROPERTY METHOD

In the simplest cases of materials selection one material property can stand out as the most critical service requirement. In such cases it is possible to estimate how much various materials to provide this requirement will cost. Cost per unit tensile strength is usually one of the most important criteria and by introducing the density and market price, the cost of buying one

MN/m² of strength (C) can be calculated as:

$$C = \frac{P \times \rho}{\sigma} \tag{13.1}$$

where P is the material price per unit weight, ρ is material density and σ is tensile strength. Eqn. (13.1) was earlier used to compare some common engineering materials, as shown in Fig. 9.5.

Equations similar to Eqn. (13.1) can be written to compare materials on the basis of modulus of elasticity, fatigue strength, creep strength, and so forth. The lower the value of C the better the material. Eqn. (13.1) can be modified to calculate the cost per unit strength and unit stiffness for different modes of loading and different materials sections, as shown in Table 13.1. In general, the formulae shown in Table 13.1 are based on the assumption that each of the materials will serve the same purpose structurally.

Since comparison between materials is a fundamental part of materials selection, a basis material can be selected and other candidate materials

TABLE 13.1

TYPICAL FORMULAE FOR ESTIMATING COST PER UNIT PROPERTY

Type of structure and loading	*Cost of unit strength*	*Cost of unit stiffness*
Solid cylinder in tension or compression	$\frac{P\rho}{\sigma}$	$\frac{P\rho}{E}$
Solid cylinder in bending	$\frac{P\rho}{\sigma^{2/3}}$	$\frac{P\rho}{E^{1/2}}$
Solid cylinder in torsion	$\frac{P\rho}{\sigma^{2/3}}$	$\frac{P\rho}{G^{1/2}}$
Solid rectangle in bending	$\frac{P\rho}{\sigma^{1/2}}$	$\frac{P\rho}{E^{1/3}}$
Solid cylindrical bars as slender columns	—	$\frac{P\rho}{E^{1/2}}$
Thin-wall cylindrical pressure vessels	$\frac{P\rho}{\sigma}$	—

σ = strength, P = price per unit weight, ρ = density, E = Young's modulus, G = shear modulus.

compared against it. The relative cost per unit property (*RC*) is given by:

$$RC = \frac{P_i}{P_b} \times \frac{\rho_i}{\rho_b} \times \frac{\sigma_b}{\sigma_i} \quad (13.2)$$

where i denotes the candidate material and b the basis material. *RC* of less than unity indicates that the candidate material is less expensive than the basis material. Equations similar to Eqn. (13.2) can be written for the different materials selections and modes of loading.

As shown in previous chapters, manufacturing costs are a significant factor in evaluating materials. This can be taken into consideration in the cost per unit property analysis by considering *P* as the cost of material on the job, i.e. the cost of stock material plus the processing and finishing costs.

Although the cost per unit property method can be useful in some cases, it has the drawback of considering one property only as the most critical and ignoring other properties. In many engineering applications, the situation is more complicated than this and material requirements invariably specify more than one property as being important.

13.3 WEIGHTED PROPERTIES METHOD

The weighted properties method can be used in evaluating complicated combinations of materials and properties. In this method each material property is assigned a certain weight, depending on its importance. A weighted property value is obtained by multiplying the numerical value of the material property by the weighting factor (α). The individual weighted property values of each material are then summed to give a comparative materials performance index (γ). The material with the highest γ is considered to be the best.

In its simple form the weighted properties method has the drawback of having to combine unlike units, which could yield irrational results. This is particularly true when different mechanical, physical and chemical properties with widely different numerical values are combined. The property with higher numerical value will have more influence than is warranted by its weighting factor. This drawback is overcome by introducing scaling factors. Each property is so scaled that its highest numerical value does not exceed 100. When evaluating a list of candidate materials, one property is considered at a time. The best value in the list is rated as 100 and the others are scaled proportionally. Introducing a scaling factor facilitates the conversion of normal material property values to scaled dimensionless values. For a given property, the scaled value (β) for a given candidate material is equal to:

$$\beta = (\text{numerical value of property}) \times 100/(\text{maximum value in the list}) \quad (13.3)$$

By this procedure, each property is given equal importance and affects the comparative materials performance index (γ) according to its weighting factor only.

For material properties that can be represented by numerical values application of the above procedure is a simple matter; but with properties like corrosion and wear resistance, service life, weldability, and so forth, numerical values are rarely given. In such cases the scaling of the material properties will have to be derived from test data and previous experience. It should be noted that in some applications, certain material properties are more desirable when they have lower numerical values as in the case of density, electrical resistivity, weight gain in oxidation and so forth. For these properties the lowest value, rather than the highest, should be rated as 100.

In cases where numerous material properties are specified and the relative importance of each property is not obvious, determinations of the weighting factors (α) can be largely intuitive which reduces the reliability of this method. This problem can be solved by adopting a systematic approach to the determination of weighting factors, and the digital logic approach, described in Chapter 8 for cost-effectiveness analysis, is an example. To increase the accuracy of decisions based on the digital logic approach, the yes–no evaluations can be modified by allocating gradation marks ranging from 0 (no difference in importance) to 3 (large difference in importance). In this case, the total gradation marks for each selection criterion are reached by adding up the individual gradation marks. The weighting factors (α) are then found by dividing these total gradation marks by their grand total.

The stock material cost and processing costs can be considered as one of the properties and given the appropriate weighting factor (α). However, in cases where the number of specified properties is large, the importance of the cost may be emphasised by considering it separately as a modifier to the material performance index (γ).

A figure of merit (*M*) for a material can be defined:

$$M = \frac{\gamma}{P \times \rho} \quad (13.4)$$

where ρ is material density, and P is the total material cost on the job, which includes cost of stock material plus the processing and finishing costs expressed as per unit weight.

When it is desired to rank a given list of materials in relation to a basis material, this can be done by computing the relative figures of merit (RM):

$$RM = \frac{M_i}{M_b} \tag{13.5}$$

where M_i and M_b are the figures of merit of the candidate material and basis material respectively.

If RM is greater than unity, the candidate material is more suitable than the basis material. In cases where M_b represents the minimum requirements, any candidate material with RM less than unity should be rejected as unsuitable.

When evaluating a large number of materials with a large number of specified properties, the above method can involve a large number of tedious calculations. Under these conditions, the use of a computer would facilitate the selection process. The steps involved in the weighted properties method can be written in the form of a simple computer program to select materials from a given data bank. In this case, the input data include the specified properties and the weighting factors (α) for the different properties. The computer can be asked to list the candidate materials in order of figures of merit, for example, or to just select the best three candidates.

13.4 INCREMENTAL RETURN METHOD

In some applications, the situation arises where a number of materials fulfil certain minimum requirements for a given application. If these materials are processed, it can be assumed that the performance levels of the resulting components are proportional to the comparative performance indices of the materials (γ). The cost of each component is also expected to vary in proportion to the materials and processing costs involved in its manufacture. If it is desired to select the component that will give the best incremental performance at the most reasonable incremental cost, the benefit–cost analysis technique described in Chapter 8 can be adopted as follows:

The material with the lowest cost (No. 1) is used as the basis and compared with the next higher cost material (No. 2). This is done by computing the incremental comparative materials performance index ($\Delta\gamma$) and the incremental cost per unit volume (P). These parameters are defined as:

$$\Delta\gamma = \gamma_2 - \gamma_1 \text{ and } \Delta P\rho = P_2\rho_2 - P_1\rho_2$$

where γ, P and ρ are as defined before and subscripts 1 and 2 refer to materials No. 1 and No. 2. If $\Delta\gamma/\Delta P\rho$ is less than unity, material No. 1 is better than No. 2, which is rejected, and the comparison proceeds between materials No. 1 and No. 3. If $\Delta\gamma/\Delta P\rho$ is greater than unity, material No. 1 is rejected and No. 2 becomes the new current best or basis material. The comparison then proceeds between materials No. 2 and No. 3. The procedure is repeated until all the alternatives have been rejected in favour of one which is considered the optimum material.

13.5 LIMITS ON PROPERTIES METHOD

In the limits on properties method of evaluation, the specified material requirements are divided into three categories: lower limit properties, upper limit properties and target values. For example, if it is desired to have a strong, light material, a lower limit on the strength properties is specified and an upper limit on the density is specified. In some cases such as for a specified failure strength or a desired thermal expansion, a target value may be specified. Whether a property is to be specified as an upper limit or a lower limit depends on the application. An example is thermal or electrical conductivities; when selecting an insulating material these properties should be upper limit specifications and when selecting a conductor they should be lower limit specifications.

In this method also, each material property is assigned a certain weighting factor (α) which can be determined using the digital logic approach, as discussed earlier. For the specified material properties, $Y_{l1}, Y_{l2}, Y_{l3}\ldots$ represent lower limits for properties 1, 2, 3 . . . ; Y_{u1}, Y_{u2}, $Y_{u3}\ldots$ represent upper limits; and Y_{t1}, Y_{t2}, $Y_{t3}\ldots$ represent target values. The corresponding properties for a candidate material j are, x_{l1j}, x_{l2j}, $x_{l3j}\ldots$; X_{u1j}, X_{u2j}, $X_{u3j}\ldots$; and X_{t1j}, X_{t2j}, $X_{t3j}\ldots$ For lower limit properties if Y_l/X_l is less than or equal to unity, the material is acceptable and for upper limit properties if X_u/Y_u is less than or equal to unity the material is also acceptable. For target value properties $|(X_t/Y_t)-1|$ has to be within the specified limit for the material to be acceptable. A candidate material which violates any of these limits is rejected. Materials which pass this preliminary screening are then graded by calculating the merit parameter (δ) which is defined as:

$$\delta = \sum_{i=1}^{nl} \alpha_{li} \frac{Y_{li}}{X_{li}} + \sum_{i=1}^{nu} \alpha_{ui} \frac{X_{ui}}{Y_{ui}} + \sum_{i=1}^{nt} \alpha_{ti} \left| \frac{X_{ti}}{Y_{ti}} - 1 \right| \qquad (13.6)$$

where *nl*, *n*u and *n*t represent the number of lower limit properties, upper limit properties and target values respectively. The lower the value of the merit parameter (δ), the better the material.

In the limits on properties method, stock material costs and processing costs can be considered in two days. The first way is to treat all cost elements as upper limit properties with the appropriate weighting factor (α). The value of α depends on the type of industry and the application for which the material is intended. For example, in consumer products the values of α are expected to be higher than those adopted in aerospace systems. This approach is based on the fact that in any engineering application there is an upper limit that can be paid for a material to meet the application requirements. However, in cases where the number of specified material properties is large, including the cost as one of the properties may obscure its importance. This is particularly true in the case of mass production industries where materials costs make up a large proportion of the total cost of the product. Under these conditions, a second approach can be adopted in considering materials and processing costs. This involves using a cost modified merit parameter Δ which is defined as:

$$\Delta = \frac{X_c}{Y_c} \times \delta \qquad (13.7)$$

where Y_c is the specified upper limit materials and processing costs and X_c is the corresponding cost of the candidate material. δ in Eqn. (13.7) is calculated as discussed earlier without taking the cost into consideration. If the value of X_c/Y_c exceeds unity the candidate material should be rejected, and the material with the lowest Δ is considered to be the most desirable material.

The limits on properties method of evaluating materials can easily be adapted for the use of a computer in selecting materials out of a data bank. As in the weighted property method, the input data include the specified properties and the weighting factors (α) for the different properties. The output can be a list of the three best candidate materials, for example, or even a complete listing of all possible materials in order of their merit.

The use of the above quantitative methods of materials selection will be illustrated for different cases in Chapters 14 to 20.

BIBLIOGRAPHY

1. P.M. APPOO and W.O. ALEXANDER, *Metals and Materials*, July/August 1976, 42.
2. H.R. CLAUSER, R.J. FABIAN and J.A. MOCK, *Materials in Design Engineering*, July 1965, 109.
3. D.P. HANLEY and E. HOBSON, *J. of Engineering Materials and Technology*, October 1973, 197.
4. *Metals Handbook,* Vol. 1, 'Properties and Selection of Metals', 8th Edn, American Soc. For Metals, 1961.

14

Selection of Materials for Cutting Tools

14.1 INTRODUCTION

Machining of materials consists of forcing a cutting tool with one or more cutting edges through the excess material on a workpiece. This excess material is progressively separated from the workpiece in the form of chips. The kind of surface produced by machining depends upon the shape of the cutting tool and/or upon the path of the tool as it traverses the material. Generally the work done in forming the chip is used to overcome the shearing forces required to separate it from the workpiece and the frictional forces it exerts against the tool faces. The forces acting on a single-point tool in orthogonal cutting can be divided into longitudinal, tangential and radial components. For a given workpiece material, these forces depend upon a number of considerations:

1. Tool forces are not changed significantly by changing the cutting speed.
2. Tool forces increase with increasing feed and depth of cut.
3. The tangential force increases with increasing chip size.
4. The longitudinal force decreases as the tool nose radius increases.
5. The tangential force decreases with increasing rake angle.
6. Using a coolant reduces the cutting forces slightly, but increases the tool life considerably by reducing friction, welding of the chip to the tool and tool tip temperature.

14.2 FACTORS AFFECTING CUTTING TOOL PERFORMANCE

Discussions in previous chapters have shown that the performance of a component in service depends on its material behaviour, its design and service conditions. Applying this analysis to cutting tools, the factors that affect their performance can be grouped as follows:

1. Material behaviour, e.g. mechanical properties of tool material, its heat treatability and depth of hard layer and its machinability and grindability.
2. Cutting tool design, e.g. single point or multiple-point, tool shape and angles.
3. Service conditions, e.g. mechanical, chemical and physical properties of the material to be machined; nature of the cutting operation and required output; condition of the machine tool and its size and power; cutting parameters, including depth of cut, feed and speed; type of coolant used.

Apart from cost, the most important tool performance considerations are tool life and cutting edge toughness. The life of a tool is an important factor in production work, since considerable time is lost whenever a tool is ground and reset. Tool life is the length of time a tool will cut satisfactorily and may be measured in a number of ways. A tool can be considered to have failed when the surface finish of the workpiece deteriorates beyond a certain limit or when the cutting forces exceed certain values. Generally, tools fail due to wear resulting from the abrading action of the metal being cut.

Since tool life decreases as the cutting speed is increased, tool-life curves are usually plotted as tool life in units of time or in units of volume of metal removed, against cutting speed. The relation between tool life (T) and cutting speed (V) can be represented as:

$$V\,T^{\mathrm{n}} = C \tag{14.1}$$

where C is a constant equal to the cutting speed for a tool life of unity. The exponent (n) depends on the characteristics of the workpiece and the tool materials as well as on the cutting conditions. For a given set of cutting conditions, longer tool life is expected with tool materials of higher abrasion resistance. For many types of cutting tool materials, higher abrasion resistance is accompanied by higher hardness both at room temperature and at high temperatures.

Besides failure due to wear, cutting tools may fail by sudden fracture of the cutting edge or by thermal spalling. Failure by sudden fracture occurs in tools made from brittle materials when taking too heavy a cut with too small a lip angle. The risk of sudden fracture can be reduced by employing tougher tool materials and rigidly supporting the tool to reduce its vibrations. Thermal spalling is usually encountered with tool materials of low thermal shock resistance. This is related to poor thermal conductivity, low ductility and a high thermal expansion coefficient.

The above interrelated variables which affect the performance of cutting tools can be correlated to measurable tool material properties as shown in Table 14.1. In this table, the first column contains the five categories of material performance requirements which were discussed in Chapter 12. The second and third columns give the corresponding tool performance requirements and the nearest measurable material properties, respectively.

14.3 ANALYSIS OF TOOL MATERIAL REQUIREMENTS

Discussions in the previous sections have shown that the major tool material

TABLE 14.1

MATERIAL REQUIREMENTS IN CUTTING TOOLS

Category	*Cutting tool performance requirement*	*Corresponding tool material properties*
Functional requirements	Tool life	Room temperature hardness High temperature hardness
	Maximum allowable cutting speed	High temperature hardness Toughness
	Edge toughness	Toughness
	Rigidity	Young's modulus
Processability requirements	Heat treatability	Hardenability (for steels) Dimensional changes (for steels)
	Possibility of intricate shapes	Formability, machinability and grindability
Cost	Cost of finished tool	Material cost Processing cost
Reliability	Probability of sudden failure and accidental breaking	Toughness Homogeneity
Resistance to service conditions	Resistance to vibrations	Toughness, Young's modulus
	Resistance to mechanical shocks	Toughness
	Resistance to thermal shocks	Expansion coefficient Thermal conductivity Ductility

functional requirements are room temperature hardness, high temperature hardness and toughness, as shown in Table 14.1. Young's modulus should also be considered in applications where tool rigidity is important, especially in cases where the tool cross sectional area and shape is limited by other geometric considerations.

Processability of the tool material affects the cost of the finished tool and can place limitations on its possible geometries. The major processability requirements are formability by casting or forging, machinability in the unhardened condition, and grindability in the hardened condition. For steels, heat treatability is also an important parameter. This includes hardenability, susceptibility to decarburisation and dimensional changes on hardening. In spite of their importance, processability requirements are difficult to quantify and in the following discussions the different materials will be given relative rating numbers based on the literature descriptions of their behaviour.

The cost of tool materials can be a relatively large fraction of the total cost of large, simple tools. However, the fraction decreases as the tools increase in complexity. Although high speed steels are less expensive than carbides, when compared on the basis of cost per unit weight, a tool made out of a solid piece of high speed steel can be more expensive than a carbide insert brazed to a plain carbon steel shank. However, because of this difference in tool construction, the high speed steel tool will permit a greater number of grinds than the carbide tool.

Service conditions have considerable effect on the performance of cutting tools. As will be shown in the following section, many of the cutting tool materials are relatively brittle which makes them unable to resist shocks or vibrations. Shocks are encountered when taking heavy cuts on rough workpiece surfaces or when the workpiece material contains hard inclusions. Interrupted cutting also leads to shocks. Vibrations are generated at the cutting edge when the tool is not rigidly supported or when the power needed for cutting approaches the maximum limits of the machine tool. The main material properties that will enable the cutting tool to resist these service conditions are: high toughness and Young's modulus for good mechanical shock and vibration resistance; and low expansion coefficient, high thermal conductivity and high ductility for good thermal shock resistance.

14.4 CLASSIFICATION OF TOOL MATERIALS

To accommodate the severe demands imposed on modern cutting tools, a wide variety of materials have been developed. The cutting tool materials most widely used for production machining can be classified into the groups given in Table 14.2.

Tool steels

The American Iron and Steel Institute (AISI) has an identification and type classification of tool steels which is based on the end use, common properties or manner of heat treatment. The AISI system of classification embraces a total of 84 different basic tool steel compositions of which a representative sample is given in Table 14.2. High speed steels retain enough hardness to cut metals at rapid rates that generate tool temperatures up to about 900 K, and yet return to their original hardness when cooled to room temperature. Because of their high content of alloy carbides, high speed steels also have greater resistance to wear than carbon or low-alloy tool steels. Even when fully hardened, high speed steels have greater toughness than carbides or cast cobalt–chromium-base alloys. However, except for short runs, carbides or cast alloys can be cheaper per piece machined in some cutting operations.

Of the different compositions four grades, T1, M1, M2 and M10, make up a considerable proportion of the tool steels used for production machining and are the most readily available. With minor variations, these four grades are closely similar in performance. Adding cobalt to high speed steels increases their cutting efficiency but also increases their price and makes them more difficult to heat treat. These superhigh speed steels are used principally for specific applications where other steels prove unsuitable.

Cast alloys

Cast cobalt–chromium–tungsten non-ferrous alloys are excellent materials for cutting tools. These alloys are cast to shape, have high red hardness and are able to maintain good cutting edges on tools at temperatures in excess of 1200 K. Compared with high speed steels, they can be used at about twice the cutting speed and still maintain the same feed. However, they are less tough, do not respond to heat treatment, and can be machined only by grinding. Because of these difficulties, cast alloys are most useful as single-point tools and are not practical for most drills, reamers, taps and broaches.

Sintered carbides

Sintered carbides are products of powder metallurgy where carbides of tungsten, titanium or tantalum are held together with a metal binder, usually

TABLE 14.2

CLASSIFICATION AND COMPOSITION OF SOME CUTTING TOOL MATERIALS

Grade	*Material*	*C*	*Mn*	*Si*	*Cr*	*V*	*W*	*Mo*	*Co*	*TaC+ TiC*	*Other*
Tool steels											
W1	Water-hardening	0·6–1·4									
W2	Water-hardening	0·6–1·4				0·25					
S1	Shock-resisting	0·50			1·50		2·50				
S2	Shock-resisting	0·50		1·00				0·50			
O1	Oil-hardening	0·90	1·00		0·50		0·50				
A2	Air-hardening	1·00			5·00			1·00			
D2	High carbon–chromium	1·50			12·00			1·00			
T1	Tungsten high speed	0·70			4·00	1·00	18·00				
T2	Tungsten high speed	0·85			4·00	2·00	18·00				
M2	Molybdenum high speed	0·85			4·00	2·00	6·25	5·00			
M3	Molybdenum high speed	1·00			4·00	2·40	6·00	5·00			
M4	Molybdenum high speed	1·30			4·00	4·00	5·50	4·50			
M10	Molybdenum high speed	0·85			4·00	2·00		8·00			
Sintered carbides											
Group 1	Sintered carbide								2·5–6·5	0–3	WC Rem.*
Group 3	Sintered carbide								15–30	0–5	WC Rem.*

(Contd.)

TABLE 14.2 (*Contd.*)

Grade	*Material*	*C*	*Mn*	*Si*	*Cr*	*V*	*W*	*Mo*	*Co*	*TaC+ TiC*	*Other*
Group 5	Sintered carbide								7–10	10–22	WCRem.*
Group 8	Sintered carbide								8–10	12–20	WCRem.*
Co–Cr–W–Mo alloys											
18% W, 2·5% C	Cast alloy (hard)	2·5	1·0	1·00	30–32		17–18·5	0·80	Rem.*		2·5–3·5 Ni
11% W, 2% C	Cast alloy (medium)	1·35–2·45	1·0	1·00	30·5–31		10–12		Rem.*		3·00 Ni
4% W, 1% C	Cast alloy (soft)	0·6–1·6	1·0–2·0	1·5–2·0	30–31		4·50	1·5	Rem.*		3·00 Ni
Ceramic											
Ceramic	Aluminium										99·9 A_2O_3

*Rem. = remainder

cobalt. Sintered carbides are usually used as inserts brazed or mechanically fastened to tool bodies, holders, or shanks.

The red hardness of carbide tools is superior to high speed steels and cast alloys and they maintain a cutting edge at temperatures of about 1470 K. However, they are very brittle and have low shock resistance which makes it necessary to provide very rigid tool support to avoid cracking. Grinding of carbides is also difficult and can be done only with silicon carbide or diamond wheels. Because of their high hot hardness, carbide tools permit cutting speeds two to three times that of cast alloy tools. Machine tools using carbides must be rigidly built, have ample power and have a suitable range of feeds and speeds.

Ceramic tools

Sintered aluminium oxide with additives of titanium, magnesium or chromium oxide can be used as cutting tool inserts either clamped onto the tool holder or bonded to it with epoxy resin. The softening point of these materials is above 1400 K which enables the tool to operate at high cutting speeds and to take deep cuts. The advantages of ceramic tools include high hardness and compressive strength at low and high temperatures and resistance to cratering. Their use is only limited by their brittleness and the machine tool condition. As in the case of carbide tools, the machine tool must be rigid, powerful and have a suitable range of feeds and speeds.

Diamonds

Diamonds can be used as a single-point tools for light cuts and high speeds but must be rigidly supported because of their brittleness. Diamonds are usually used for special purpose cutting and usually give longer life than carbides. They are also used for dressing grinding wheels.

14.5 ALLOCATION OF WEIGHTING FACTORS

Previous discussions have shown that numerous material requirements are involved when selecting tool materials. These material requirements are:

1. Room temperature hardness.
2. Hardness at operating temperature.
3. Toughness.
4. Ductility.
5. Thermal conductivity.

6. Thermal expansion coefficient.
7. Young's modulus.
8. Machinability and grindability.
9. Heat treatability for steels.
10. Availability.
11. Cost (stock material cost + processing cost).
12. Reliability (not subject to sudden failure and accidental breaking).
13. Need for special machine tools.

The first three requirements are of major importance for all classes of cutting tool materials as they determine the tool life and rate of metal removal, and their values are given in Table 14.3 for the different tool materials. Although ductility is less important than the first three requirements, brittle tool materials can crack during setting up if not carefully handled or correctly mounted. This means that longer setting up time should be allowed for brittle tools. Ductility can also be taken as a measure of reliability as defined in requirement 12. Brittle cutting tool materials are also prone to failure by thermal shock especially when their thermal conductivity is low and their thermal expansion coefficient is high. Requirements 4, 5 and 6 should be considered for brittle tool materials, especially when comparing ceramics with other materials. As ductility, thermal conductivity, thermal expansion coefficient and reliability are closely related and are only of interest for very brittle materials, they will be grouped together as one parameter, liability to fracture. This will be given a rating on a scale from 1 to 9 with lower values for brittle materials that are difficult to handle and prone to thermal shock and higher values for ductile conducting materials, as shown in Table 14.3.

The value of Young's modulus determines the rigidity of a tool of a given geometry. Although carbides, cast alloys and ceramics have higher moduli than steels, they are mostly used as inserts on steel holders so that the effective rigidity is that of the steel. Under these conditions there is no need to include the Young's modulus as a parameter in the selection process.

Machinability and grindability requirements can be included under the cost parameter as parts of processing costs. Differences in heat treatability of tools steels can also be accounted for in processing costs in the case of simple tools. When tool geometry is intricate and dimensional tolerances are narrow, heat treatability should be considered as a go–no–go parameter to exclude steels that cannot possibly be manufactured according to requirements. This situation usually arises when selecting from high speed steels alone and is too specific to discuss in this case study.

TABLE 14.3

PROPERTIES OF CUTTING TOOL MATERIALS

Grade	*Room temperature hardness (Rc)*	*Hardness at 380 K (Rc)*	*Toughness (Joule)*	*Liability to fracture 1–9**	*Cost index***
Tool steels					
W1	63	10	68	8	100
W2	63	10	68	8	100
S1	60	20	95	9	100
S2	63	20	95	9	100
O1	63	20	54	8	100
A2	63	30	48	7	100
D2	62	35	30	7	98
T1	66	52	61	8	84
T2	65	52	61	8	84
M2	65	52	68	9	88
M3	67	52	48	8	86
M4	67	52	48	8	86
M10	65	52	68	9	92
Sintered carbides					
Group 1	76	69	0·97	3	29
Group 3	73	65	2·4	4	31
Group 5	73	65	0·8	2	36
Group 8	75	69	0·8	2	32
Co–Cr–W–Mo alloys					
18% W, 2·5% C	62	52	3·4	3	61
11% W, 2% C	53	43	4·8	4	62
4% W, 1% C	41	32	12	5	63
Ceramic					
Alumina	80	72	0·7	1	80

*See text for definition of liability to fracture.
**Cost index is calculated on the basis of 'cost on the job' per unit volume. The cheapest material is given a cost index =100.

Availability of tool material is an indication of the number of different sizes available in stock, the number of warehouses that stock the material and the delivery time. Generally, materials usable for general purpose machining operations are easily available while the less popular grades are usually more expensive and have longer delivery times. For simplicity this parameter can be considered as part of the cost of stock material.

As discussed earlier, some tool materials can only be used to their best advantage if suitable machine tools are available; otherwise their use will be uneconomical. Therefore, this parameter should be treated as a go–no–go condition, depending on the available machines and the type of job.

Based on the above analysis, the weighting factors are allocated to the different requirements as shown in Table 14.4, the digital logic method having been used to estimate the weighting factors, as discussed in Chapter 8. Two different cutting conditions are considered. The first case in one where the material to be machined has a rough surface and contains hard inclusions which would subject the cutting edge to shocks so that toughness is the most important property, followed by hot hardness. The second case to be considered is that of a cutting process where a high rate of metal removal is most important, so that hot hardness is the most important property, followed by tool life as measured by room temperature and high temperature hardness.

TABLE 14.4

WEIGHTING FACTORS FOR DIFFERENT PROPERTIES UNDER DIFFERENT CUTTING CONDITIONS

Property	*Workpiece with rough surface and hard inclusions*	*High rate of metal removal with homogeneous workpiece material*
Room temperature hardness	0·15	0·25
High temperature hardness	0·25	0·40
Toughness	0·40	0·15
Liability to fracture	0·10	0·10
Cost	0·10	0·10

14.6 EVALUATION OF CUTTING TOOL MATERIALS

The cutting tool materials listed in Table 14.2 were evaluated using the weighted properties method which was discussed in Chapter 13. The most important tool material properties were shown earlier to be the room temperature hardness, high temperature hardness and toughness. In actual service, the cutting tool performance is also affected by the properties of the workpiece. For example, a cutting tool with a hardness of Rc 60 is expected to perform satisfactorily when cutting a workpiece of Rc 20, but not when cutting a workpiece of Rc 45.

In the present case study, it is assumed that the workpiece material properties are such that the minimum acceptable cutting tool material hardness is Rc 60 at room temperature and Rc 10 at 830 K. The candidate materials of Table 14.2 were rated by taking room temperature hardness of Rc 60 as unity and Rc 80 as 100. Intermediate values were rated in proportion. High temperature hardness values were treated similarly but RC 10 was taken as unity and Rc 72 as 100.

Table 14.5 lists the relative performance index (γ_R) for the candidate

TABLE 14.5

EVALUATION AND GRADING OF CUTTING TOOL MATERIALS

Tool material	*Rough surface and inclusions*		*High rate of metal removal*	
	γ_R	*preference*	γ_R	*preference*
Tool steels				
W1	75	10	44·7	16
W2	75	10	44·7	16
S1	89	7	53·5	12
S2	92	6	57·6	9
O1	75	10	50·6	13
A2	75	10	55·9	10
D2	69	13	55·5	11
T1	95·7	4	80·3	7
T2	94·7	5	78·7	8
M2	94·4	2	81·4	6
M3	97	3	82·2	4
M4	97	3	82·2	4
M10	100	1	81·9	5
Sintered carbides				
Group 1	76	9	87·9	2
Group 3	72·7	11	81·4	6
Group 5	69·3	12	78·7	8
Group 8	57·7	14	85	3
Co–Cr–W–Mo alloys				
18% W, 2·5% C	57	15	55·9	10
11% W, 2% C	52·5	16	47·5	14
4% W, 1% C	51	17	46·2	15
Ceramics				
Alumina	85	8	100	1

materials under the two cutting conditions. γ_R was calculated as:

$$\gamma_R = \frac{\gamma_i \times 100}{\gamma_{max}} \tag{14.2}$$

where γ_i is the performance index for candidate i calculated according to Section 13.3, and γ_{max} is the performance index of the best material.

For the case where the workpiece material was assumed to have a rough surface and to contain hard inclusions, the high speed steels exhibited the highest performance index. The most favourable material is M10 which has the highest toughness and lowest cost among high speed steels. M2 was found to be second best. These two materials are the most popular grades of HSS. When high rates of metal removal were required, ceramics exhibited the highest performance index followed by group 1 and 8 sintered carbides as second and third best respectively. If available machine tools cannot accommodate these materials, HSS M3, M4 and M10 can be used.

BIBLIOGRAPHY

1. *Metals Handbook*, Vol. 1, 'Properties and Selection of Metals', 8th Edn, American Soc. for Metals, 1961.
2. G.A. Roberts, J.C. Hamaker and A.R. Jonson, *Tool Steels*, American Soc. for Metals, 1962.

15

Selection of Materials for Lubricated Journal Bearings

15.1 INTRODUCTION

A bearing is a machine element which transmits loads or reaction forces from a moving part to the bearing support. Journal bearings are cylindrical and are used when the transmitted load is essentially at right angles to the axis of the rotating shaft. Besides carrying these loads, the bearing material is also subjected to the sliding movement of the shaft. The main problem in selecting bearing materials is to combine the roles of load carrying and sliding. In order to reduce the friction that results from this sliding movement, the bearings are lubricated with oil or grease. However, even in the presence of the lubricant, some direct contact between the shaft and bearing surface is unavoidable so the bearing qualities of the bearing material are important. When excessive direct contact between the shaft and bearing occurs, seizure can take place leading to bearing failure.

Generally, the performance of the bearing material is system dependent and is not only affected by the material properties but also by the lubricant and journal material. The main parameters which affect the operation of a bearing system are load, speed, temperature, degree of aeration and cleanliness. These parameters will be discussed in detail in the following section.

15.2 FACTORS AFFECTING THE PERFORMANCE OF BEARING MATERIALS

As bearings are used under widely different conditions, a bearing alloy that is well suited for one application may be unsuitable for another. The magnitude and nature of the load on the bearing are perhaps the most important service conditions that can affect the behaviour of bearing materials. Heavy loads require high compressive strength to avoid deformation of the bearing material in service. Loading conditions can vary from steady loading in one direction to cyclic loading of variable direction. Generally, variations in the loading direction are less important than variations in the magnitude of loading. The load carrying capacity under steady loading is affected by the compressive strength of the bearing material while the capacity under cyclic loading is usually determined by its fatigue strength. If the bearing length is increased to reduce the stress, alignment of the shaft may become poor. A low modulus of elasticity and large ductility are then needed to avoid high unit pressure at the bearing edges.

Temperature is another important service condition that can affect the performance of bearing materials. The operating temperature of a bearing is affected by the surface speed of the shaft, the rate of lubricant supply, the conductivity of the bearing material, and the frequency of starting and stopping. Both the strength and corrosion rate of the bearing material depend on the operating temperature.

Acids contained, or formed, in the lubricating oils during operation are the main sources of corrosion of the bearing material. Corrosion can also be caused by glavanic action between the shaft and bearing materials.

The performance of bearing materials depends on the presence of the lubricant in the bearing. The reliability of the oil supply and the amount of oil present in the bearing depend mainly on the design of the bearing. When the risk of lubricant failure is high, materials must be chosen for good resistance to scoring or seizing. Even with the best lubricant supply, metal-to-metal contact of bearing and shaft will be caused by stopping, starting, high deflections, sudden shock loads or overloads. The ability of the oil to wet the bearing material surface, called oiliness, tends to reduce scoring.

Service conditions where dirt and foreign materials can enter the bearing can affect the performance of bearing materials. A foreign particle in the bearing either cuts a groove around the bearing and shaft, or becomes embedded in the bearing material. The latter case does less harm to the bearing system and a soft bearing material is more suitable under these conditions.

Usually the shafts are more expensive to replace and wear should be concentrated in the bearing material. To ensure this, the bearing material must be considerably softer than the shaft. The shaft surface finish is an important parameter in deciding the bearing service life, and in some cases of heavily leaded, copper-base alloys, early failures can be attributed to rough shaft surfaces.

Because of the difference in thermal expansion of the shaft and bearing material, the amount of clearance changes with temperature. The bearing service temperature depends on the load and velocity. The design clearance should be calculated at the running temperature rather than at room temperature.

15.3 ANALYSIS OF BEARING MATERIAL REQUIREMENTS

The above discussions have shown that a bearing material has to have certain functional requirements. Compressive strength is important when the load is heavy enough to cause extrusion of the bearing material. Bonding the bearing material to a strong backing increases the effective strength especially for thin layers. High fatigue life of a bearing material is likely to be the most important requirement under conditions of alternating load. The fatigue life of most bearing materials increases with decreasing operating temperature and thickness. Hardness becomes important under conditions where the oil film may break down. Soft bearing materials will ensure that no wear takes place in the shaft material. Many bearings support shafts that deflect under load. This deflection causes heavy pressure at the ends of the bearing and leads to breakdown of the lubricating oil film. If the bearing material yields sufficiently to take care of the misalignment, trouble can be avoided. Materials with lower Young's modulus will undergo larger deflections under lower loads.

Apart from the above mechanical requirements, bearing materials should have good thermal conductivity. This is particularly true under conditions of high energy dissipation rate, such as high speeds or high loads. The ability of the bearing material to dissipate heat away from friction surfaces reduces the operating temperature, and thus reduces the possibility of melting the bearing material and of seizure of the bearing. Wear resistance of a bearing material can be an important parameter in cases where the position of the journal with respect to the bushing is to be kept constant within narrow tolerances, or if the bearing material is a thin layer as in the case of thin lining over a backing material. Generally, the wear rate depends on the tendency of the system towards adhesive weld formation, and on the resistance of the bearing material against abrasion by the asperities on the mating surface. The rate of wear can be an important factor in determining the bearing life.

Processing of bearing materials is not usually difficult, and the most important requirement is a good bond with the backing material. This is because most bearing materials are used in conjunction with a strong backing material, usually steel. A badly bonded bearing material can peel from its backing. However, all commercial bearing materials can be well bonded to appropriate backings provided the correct procedure is followed, and this parameter will not be considered as a selection criterion. Melting point is another processability consideration. Materials with lower melting points are easier and less expensive to cast, while higher melting point materials usually require specialised and more expensive techniques.

The cost of the bearing material stock can be a relatively high proportion of its cost on the job, especially in lower melting point materials where processing and final machining are simple. Higher melting point materials are more expensive to process and the stock material may represent a smaller proportion of the total cost. In the following analysis, the cost of the stock material and the processing costs will be considered as one item.

The main service requirements for bearing materials are corrosion resistance and embeddability. Corrosion of the bearing material causes it to thin down which increases the clearance and could lead to complete failure. Only corrosion resistant materials should be used in bearings where acidic oils may be present. The rate of corrosion increases with increasing operating temperature. This parameter is difficult to measure quantitatively and materials are usually graded as good, moderate and poor. Embeddability is the ability of the bearing material to accommodate foreign materials to avoid scoring of the shaft. This parameter becomes important under service conditions where dirt is present. The nearest mechanical property to embeddability is hardness. Lower hardness corresponds to higher embeddability.

15.4 CLASSIFICATION OF BEARING MATERIALS

A wide variety of bearing materials have been developed to accommodate the conflicting requirements of strength and sliding properties. The most widely used bearing materials can be classified into the groups given in Table 15.1.

Whitemetals (babbitt alloys)

Babbitts are tin-base or lead-base alloys with additions of antimony and

TABLE 15.1

COMPOSITION OF SOME METALLIC BEARING MATERIALS (%)

Grade	*Alloy*	*Sn*	*Sb*	*Pb*	*Cu*	*Fe*	*As (max)*	*Zn*	*Al*	*Others*
Whitemetals										
1	ASTM B23 (Tin-base)	91	4·5	0·35	4·5	0·8	0·1	0·005	0·005	0·08 Bi
2		89	7·5	0·35	3·5	0·08	0·1	0·005	0·005	0·08 Bi
3		84	8	0·35	8	0·08	0·1	0·005	0·005	0·08 Bi
4		75	12	10	3	0·08	0·15	0·005	0·005	
5		65	15	18	2	0·08	0·15	0·005	0·005	
6	ASTM 23 (Lead-base)	20	15	63·5	1·5	0·08	0·15			
7		10	15	75	0·5	0·1	0·6			
8		5	15	80	0·5		0·2			
10		2	15	83	0·5		0·2			
11			15	Rem.*	0·5		0·25			
15		1	15	Rem.*	0·5		1·4			

(*Contd.*)

TABLE 15.1 (*Contd.*)

Grade	*Alloy*	*Sn*	*Sb*	*Pb*	*Cu*	*Fe*	*As (max)*	*Zn*	*Al*	*Others*
Copper-base alloys										
48	SAE Copper– lead	0·25		25–32	67–74	0·35		0·1		1·5 Ag, 0·025 P
49		0·5		21–27	73–79	0·35				
480		0·5		30–40	60–70	0·35				1·5 Ag
A	ASTM B22 (Bronze)	18–20		0·25	79–82	0·25		0·25		1 P
B		15–17		0·25	82–85	0·25		0·25		1 P
C		9–11		8–11	78–82	0·15		0·75		0·1 P, 1 Ni
E		0·2		0·2	60–68	2–4		Rem.*		2·5–5 P, 3–7·5 Ni
2A	ASTM B143 (Leaded tin-bronze)	6		1–2	86–90	0·25		3–5		0·05 P, 1 Ni
Aluminium-base alloys										
770		5·5–7			0·7–1·3	0·7			Rem.*	0·7–1·3 Ni, 0·7 Si
780		5·5–7			0·7–1·3	0·7			Rem.*	0·3–0·7 Ni, 1–2 Si
MB7		6·5–7·5			0·7–1·3	0·6			Rem.*	1·5–1·8 Ni, 0·6 Si

*Rem. = remainder.

copper, as shown in Table 15.1. Compared with other bearing materials, whitemetals have relatively low fatigue strength which limits their use to low-load conditions. However, supporting these alloys by steel or bronze backing shells increases their load carrying capacity; and reducing their thickness increases their fatigue strength. These alloys are easy to handle commercially, and they have excellent resistance to seizure and can be used at high sliding velocities under light loads. Tin-base alloys resist corrosion much better than lead-base bearing alloys. Whitemetals are tolerant of misalignment, dirt and temporary lubrication failure. Being soft themselves, they run satisfactorily against a journal of relatively low hardness.

Whitemetals are usually manufactured by casting continuously onto prepared cold rolled steel strips and then forming the resulting bimetal into the required shape by pressing. Another method is individual casting on preformed backing material. When centrifugal casting is used, dense linings and sound bonds are obtained.

Copper-base bearing alloys

Compared with whitemetals, copper-base alloys offer a much wider range of strength and hardness, as shown in Table 15.2. Generally, the softest alloy that will provide adequate strength should be selected for a given application.

The properties of copper–lead alloys depend on the lead content, and the higher the lead content, the lower the fatigue strength but the better the sliding properties. Silver and tin may be substituted for part of the copper in order to increase fatigue strength. These alloys are easily corroded by certain oils, because lead dissolves in certain organic acids which form when oil is oxidised. This inherent disadvantage of copper–lead alloys can be overcome, at least partially, by the use of electroplated soft metal overlays. Lead–tin, lead–tin–copper, and lead–indium can be used as overlay materials. However, these overlays lose their corrosion protection in service at high temperatures owing to diffusion.

As with whitemetals, copper–lead alloys can be manufactured by casting the alloy on steel backing and then forming the bimetal to shape. The bimetal strip can also be prepared by sintering copper and lead powders on the steel strip in a reducing atmosphere. Individual lining of formed steel backing can also be used. Bronzes containing 5 to 20% tin and a small percentage of phosphorus are also used as bearing materials. These alloys are usually used where load-carrying capacity at low speeds is of paramount importance. Substitution of zinc for tin in these alloys improves their ductility but reduces their strength and hardness. In applications demanding

TABLE 15.2

ROOM TEMPERATURE PROPERTIES OF SOME METALLIC BEARING MATERIALS

Grade	*Hardness (BHN)*	*Yield point (MN/m²)*	*Ultimate strength*	*Fatigue strength (MN/m²)*	*Corrosion resistance*	*Wear resistance*	*Relative cost*	*Specific gravity*	*Thermal conductivity**	*Modulus of elasticity (GN/m²)*
Whitemetals										
1	17	30	90	26·6	5	2	4 000	7·34	0·12	51·1
2	24·5	42·7	104·3	33·6	5	2	4 000	7·39	0·12	53·2
3	27·0	46·2	123·2	37·0	5	2	4 000	7·46	0·12	53
4	24·5	38·9	113·0	31·0	5	2	4 000	7·53	0·12	53
5	22·5	35·4	105·4	28·0	5	2	4 000	7·75	0·12	53
6	21·0	26·6	101·9	22·0	4	3	700	9·6	0·057	29·4
7	22·5	24·9	109·6	28·0	4	3	650	9·73	0·057	29·4
8	20·0	23·8	109·2	27·3	4	3	600	10·04	0·058	29·4
10	17·5	23·5	108·2	27·0	4	3	550	10·0	0·058	29·4
11	15·0	21·4	89·6	22·0	4	3	550	9·9	0·058	29·4
15	21·0	28·0	72·5	30·1	4	3	550	10·1	0·058	29·4

Copper-base alloys										
48	22·5–32·5	40	56–72	45	3	5	800	9·4	0·1	75
49	30–40	45	60–80	50	3	5	800	9·35	0·1	75
480	20–32·5	38	52–63	42	3	5	800	9·5	0·1	75
A	100	168	350	120	2	5	1 000	8·8	0·1	94·5
B	100	126	280	100	2	5	1 000	8·8	0·1	94·5
C	65	119	266	91	2	3	900	8·95	0·11	77
E	223	420	770	300	2	4	850	8·75	0·1	94·5
2A	68	112	238	77	2	3	900	8·8	0·1	94·5
Aluminium-base alloys										
770	70	172·5	190	150	3	2	800	2·88	0·4	72·8
780	68	157·5	182	135	3	2	800	2·88	0·4	72·8
MB7	73	192·5	203	170	3	2	800	2·91	0·4	73·5

*Units in cal/cm^2/cm/°C/s.

even higher mechanical strength than tin-bronzes, silicon-, manganese- or aluminium-bronzes are used. The methods for making steel backed bronze bimetals are similar to those used for copper–lead alloys.

Aluminium-base alloys
Several aluminium alloys have mechanical and physical characteristics suitable for bearing applications. Their hardness and strength are suitable for high-duty bearings and their corrosion resistance is excellent. Their main disadvantage is their liability to seizure against other surfaces. This drawback can be mitigated by adding a soft metal like tin which smears over the alloy surface and prevents cold welding. However, concentrations of tin greater than 7% reduce the fatigue strength and ductility.

Steel backed aluminium bearings are manufactured by hot rolling the two materials to allow solid-phase welding. A pure aluminium foil can be introduced at the interface to enhance bonding.

Non-metallic bearing materials
Polymers can be used as bearing materials under conditions of light loading. The main advantage of polymers is their low coefficient of friction. Their low elastic modulus assists conformation to the contour of the shaft. On the other hand, polymers have low thermal conductivity and are liable to dimensional changes. The low melting temperature of nylon limits its service temperature to below 150° C, but Teflon can be used at temperatures of about 250° C. When used as bearing materials, polymers should be adequately cooled and comparatively large clearances allowed. To simplify the analysis, non-metallic bearing materials will not be considered in the following discussions.

15.5 ALLOCATION OF WEIGHTING FACTORS

The above discussions have shown that numerous material requirements are involved in the selection of bearing materials. These requirements are:

1. Hardness.
2. Yield strength.
3. Ultimate strength.
4. Fatigue strength.
5. Young's modulus.
6. Corrosion resistance.

7. Wear resistance.
8. Thermal conductivity.
9. Cost.

In the above list, the first five requirements are mechanical properties. The required values of these properties are determined from the design criteria. The hardness and Young's modulus are upper limit requirements while the yield, ultimate and fatigue strengths are lower limit requirements, using the definitions of Section 13.5. The relative importance of the five mechanical properties is determined from service conditions, e.g. under fluctuating loads the fatigue strength becomes more important while under heavy, steady loads the yield strength becomes more important. Although the required strength properties are system independent, at least to a first approximation, the hardness is system dependent and should be determined in relation to the shaft material hardness.

The corrosion resistance requirement is system dependent as it depends on the type of lubricant used, and on the possibility of galvanic action with the shaft material. The wear rate is also system dependent and is affected by the surface roughness of the shaft, lubrication and cleanliness of service conditions.

Under conditions of high speeds and heavy loads, heat is generated at high rates and could raise the bearing temperature, leading to thermal degradation and oxidation of the lubricant. Under these conditions, thermal conductivity of the bearing material becomes an important parameter.

For the present case study, two different service conditions are considered. The first case is for a heavy, fluctuating load, slow speed of rotation and frequent stopping and starting. Adequate sealing of the bearing ensures that the lubricant will not become contaminated with abrasive materials. Lubricant will not be changed frequently, and thus possible corrosion effects, by oxidation products in the lubricant, must be taken into account. The positioning function of the bearing is not important. Cost of the bearing material is not very important, as the bearing is serving in expensive equipment. The second case is for a light, steady load with high speed of rotation. The bearing is protected from abrasive materials but could be subjected to corrosive effects from the lubricant. The positioning function of the bearing is important in view of the high speeds. The bearing is to be produced as cheaply as possible and should have long service life. In both cases, the hardness of the shaft is high and can tolerate any bearing material up to phosphor bronzes.

Based on the above analysis and design considerations the weighting

TABLE 15.3

WEIGHTING FACTORS FOR DIFFERENT SERVICE CONDITIONS

Property	*Heavy, fluctuating load, slow speeds*	*Light, steady load, high speeds*
Hardness	0·10	0·10
Yield strength	0·12	0·07
Ultimate strength	0·12	0·07
Fatigue strength	0·20	0·06
Young's modulus	0·11	0·15
Corrosion resistance	0·10	0·10
Wear resistance	0·08	0·15
Thermal conductivity	0·10	0·10
Cost	0·07	0·20

factors were estimated with the help of the digital logic analysis described in Chapter 8. Table 15.3 lists the different weighting factors for the two service conditions.

15.6 EVALUATION OF BEARING MATERIALS

Evaluation of the bearing materials listed in Table 15.1 was carried out on the basis of their properties, listed in Table 15.2, and the weighting factors, listed in Table 15.3. The quantitative method used is the weighted properties method which was described in Section 13.3. The hardness, Young's modulus and cost are taken as upper limit properties while yield, ultimate and fatigue strengths are taken as lower limit properties. Corrosion resistance, wear resistance and thermal conductivity are also taken as lower limit properties. In the case of upper limit properties the least value for each property in Table 15.2 is considered the best and rated 100; while in the case of lower limit properties the highest value for each property is considered the best and rated as 100. The results of evaluation are given in Table 15.4 together with the order of preference of the different materials.

For the case of heavy, fluctuating load, case 1, grade E phosphor bronze is most favourable, followed by the aluminium alloys and then the lead babbitts, as shown in Table 15.4. This is in accordance with industrial practice. The phosphor bronze combines the best mechanical properties with reasonable cost and wear resistance. If the shaft material is too soft to tolerate phosphor bronze or aluminium alloys, lead babbitts can be used.

For the case of light, steady loads, case 2, aluminium alloys are best,

TABLE 15.4

EVALUATION AND GRADING OF BEARING MATERIALS

Grade	*Heavy, fluctuating load, slow speeds*		*Light, steady load, high speeds*	
	γ	*Preference*	γ	*Preference*
Whitemetals				
1	35·8	11	39·6	14
2	34·1	14	37·4	18
3	34·2	13	37·1	19
4	34·1	14	37·4	18
5	34·3	12	37·8	16
6	38·5	10	48·2	10
7	39·8	7	48·3	9
8	39·5	8	49·5	8
10	41·0	6	51·6	5
11	41·8	5	52·7	4
15	39·3	9	49·9	7
Copper-base alloys				
48	33·4	16	42·9	11
49	33·0	18	42·3	12
480	33·2	17	42·9	11
A	39·5	8	41·1	13
B	35·8	11	39·3	15
C	33·7	15	34·6	17
E	63·2	1	50·0	6
2A	30·9	19	32·6	20
Aluminium-base alloys				
770	50·6	3	57·7	2
780	49·1	4	57·2	3
MB7	52·6	2	58·4	1

followed by grades 11 and 10 of the lead babbitts. The high conductivity and low cost of aluminium alloys are their main attractions.

BIBLIOGRAPHY

1. A.W.J. DE GEE, *Wear*, **36**(1976), 33.
2. P.G. FORRESTER, 'Materials for Plain Bearings', *Modern Materials*, Vol. 4, Academic Press (New York, London), 1964, p. 173.
3. P.G. FORRESTER, 'Selection of Plain Bearing Materials', *Engineering Materials, Selection and Value Analysis*, H.J. Sharp, Ed., Heywood (London), 1966, p. 255.

16

Selection of Materials for Electrical Insulation

16.1 INTRODUCTION

Generally, electrical insulating materials offer a very high resistance to flow of electric current so that they can be used to confine the current in its proper path along the conductor. Insulators may be solid, liquid or gaseous materials. Besides the primary function of electrical insulation, thermal, mechanical and chemical functions may have to be performed by the insulator material. Often the secondary requirements are so necessary that they, rather than the electrical properties, limit the choice of material for a given application, e.g. heat resistance is essential in heating devices and flexibility is essential for insulating wires and cables.

16.2 FUNCTIONS OF INSULATORS

The functions of insulators are diverse and depend on the requirements, design and service conditions. Separation of conductors from ground and from each other is the classical requirement of insulators, i.e. spacing insulation. Barrier insulation is required to develop higher dielectric strength than that of separation type, and creepage insulation is required to resist flashover. Creepage resistance is affected by surface contamination and moisture. High voltage isolation must either be little affected by corona or be used in corona-free devices. A prime requisite for solid insulators under high voltage conditions is dense internal structure not containing internal voids.

Besides the above electrical functions, insulators may have to perform mechanical functions like supporting the conductor or resisting centrifugal forces which require high mechanical strength. If the conductor is to be wound, the insulator should also be flexible. Hardness and modulus of elasticity may also be important for applications where abrasion resistance and rigidity are required respectively. Another secondary function of insulating materials is protection of conductors from contamination, abrasion and chemical attack. In some special applications, the electrical insulation has to contend with radiation effects, shocks and vibrations.

16.3 ANALYSIS OF INSULATION MATERIAL REQUIREMENTS

Functional requirements

The functional requirements of insulating materials are, by necessity, all electrical in nature. These include dielectric strength, dielectric constant, dielectric loss, resistivity, flashover and tracking. The behaviour of insulating materials in terms of these requirements has an important bearing on their selection, as will be briefly discussed here.

The basic property required for insulation is dielectric strength which is defined as the voltage at which breakdown of the insulation occurs. This is dependent on inherent and imposed factors such as:

1. Intrinsic breakdown strength of the material under special test conditions.
2. The moisture content of an insulating material or system plays an important part in determining its breakdown strength. For a given material, the wet breakdown strength may be about 50% lower than the dry strength.
3. Insulation thickness is an important factor in determining the dielectric strength of thin films. The breakdown voltage rises linearly in a log–log plot of voltage gradient versus thickness with thicknesses less than about 2·5 mm.
4. The surrounding medium can affect the dielectric strength of a material or system, e.g. conducting media must be avoided to ensure adequate dielectric strength levels.
5. Increasing the temperature usually reduces the dielectric strength and some materials exhibit linear relations between log dielectric strength and reciprocal of absolute temperature.

These should be taken into account when considering the value of dielectric strength for a given application. The intrinsic breakdown strength, however, is useful for comparing different materials.

The dielectric constant of insulating materials is a measure of electrostatic energy stored in the material per unit volume per unit voltage gradient. If the insulating material is to be used in a capacitor, a high dielectric constant is important. Conversely, high-voltage cable requirements call for high dielectric strength but a very low dielectric constant. For small d.c. motors and small power transformers for low frequency electronic power supplies, the dielectric constant is of little importance. When insulating materials are used at high frequencies, the power loss in the material itself may be very important. The power loss varies directly with the dielectric constant and the dissipation factor. The dissipation factor is defined as $\tan \delta = R_p/2f C_p$ where C_p = equivalent parallel capacitance of the insulating material, R_p = equivalent parallel resistance of the insulating material, and f = frequency in cycles/sec. This definition of the dissipation factor is based on considering the material as a combination of resistance and pure capacitance.

The insulation resistance of an insulator depends on both the properties of the material and the geometry of the insulator, and it is usual to consider surface and volume resistivity of the material. Surface resistivity can be described as the resistance between two opposite edges of a square portion of the insulating surface, and volume resistivity is the resistance between any two opposite faces of a unit cube of the insulating material. With inhomogeneous materials, the surface and volume resistivity may have different values depending on orientation. Volume resistivity is often affected by surface resistivity, thus in designing, it is important to provide sufficient surface creepage distance to be able to use materials with high volume resistivity levels. Insufficient creepage distance and contamination can lead to either flashover or tracking. Resistivity is affected by external parameters in a similar fashion to dielectric strength.

Flashover and tracking are surface phenomena. Flashover is the breakdown of the air or gas in parallel with solid or liquid insulation. Tracking is a disruption of a solid insulation surface between two electrodes. The latter results in a carbonised path which eventually shorts the electrodes. Flashover is controlled by creepage distance and insulator surface design, and the flashover voltage is affected by humidity, type of voltage, contamination and dielectric constant of solid insulators.

Environmental factors

Insulation breakdowns are frequently caused by heat and moisture. Thermal ageing usually reduces the mechanical strength of insulating materials and increases with increasing temperature.

Insulating materials must sometimes serve as heat conductors. The internal temperature of a winding depends upon the rate at which heat is conducted through the insulation to the surroundings. Where an insulation is subjected to wide temperature changes, the thermal expansion coefficient may become a limiting factor.

Another form of ageing is that caused by electromagnetic radiation, e.g. gamma rays and X-rays, and nuclear radiation, e.g. alpha particles, electrons neutrons, charged ions and molecules. This form of ageing usually affects the mechanical properties, and can also reduce insulation resistance and dielectric strength.

Mechanical properties of insulating materials can be an important factor in determining their service life. Fatigue, creep and stress cracking can occur in service. Fatigue can occur in a similar way to metal fatigue and is usually speeded up by thermal degradation. Creep can cause separation between insulator and conductor which could impair heat conduction and lead to eventual failure of the system. Stress cracking is often associated with other environmental conditions such as radiation, corona and thermal ageing.

The life of insulating materials can also be affected by water, oil, ozone, corrosive vapours, acids, alkalies and other chemicals. As discussed before, water can directly affect the electrical properties of the insulation while oil and ozone can degrade organic insulators. Corrosive vapours can attack the conductor material after penetrating the insulator. This may impair the effectiveness of the adhesion between the conductor and the insulating coating, thus leading to deterioration of the latter, and even possible failure.

Having considered the different aspects of environmental factors that can affect the performance of insulating materials, it is important to bear in mind that almost always two or more environmental effects will reinforce each other's separate effects and speed up the final breakdown of insulation.

Processability requirements

Besides the usual processability requirements, e.g. ability to be shaped in the required form at reasonable cost, insulating materials must also meet the requirement of withstanding the hazards of final processing. In winding electrical tapes, for example, winding tension must not injure the material. Wound coils may have to be pushed into motor slots, formed by bending, crushed to conform to existing space or twisted to form desired shapes. These operations require a high level of mechanical properties in insulating materials to assure freedom from cracks.

After forming, the component may have to be dried under high heat and in vacuum. These processing conditions may cause shrinkage cracks, heat shock, internal gas evolution and embrittlement. These hazards can be minimised by considering processability requirements during the selection process.

16.4 CLASSIFICATION OF INSULATING MATERIALS

The materials ordinarily employed in insulation include: solids such as cotton, silk, paper, rubber, wax, resins, asphalts, varnishes, polymers, mica, glass, asbestos, porcelain and pure oxides; dielectric liquids such as petroleum oils and synthetic oils; and dielectric gases such as hydrogen, dichlorodifluoromethane, nitrogen and air. This classification gives the state in which the materials are used under service conditions but does not indicate their service limits. A more useful classification is given according to the permissible operating temperature as shown in Table 16.1. This classification was established by the Institute of Electrical and Electronic Engineers (IEEE) for solid insulating materials and the limiting temperature is assumed to be at the hottest point. The form of the insulating material greatly affects its performance and in many cases some materials are only available in a certain form, e.g. papers, tapes and fabrics are only available as films, while ceramics are usually available in bulk or tube form. This form limitation is an important criterion in the selection of insulating materials. The following discussion will be mainly concerned with materials that are usually used in bulk electrical insulation. This class of materials is not only applied in bulk form but is also usually produced in bulk quantities. The

TABLE 16.1

IEEE THERMAL CLASSIFICATION SYSTEM

Class	*Temperature (K)*	*Material*
O	363	Unimpregnated cellulose, cotton, silk, paper
A	378	Polyvinyl acetal, polyurethane, polyamide and class O materials in the impregnated form
B	403	Epoxy, polyester-glass, mica, asbestos
F	428	Modified silicone, terephthalate polyester
H	453	Silicone, silicone rubber, esterimide
K or 200	473	Amide-imide overcoated ester-imide
M or 220	493	Amide-imide, polyimide
C	523 +	Polytetrafluoroethylene, glass, ceramics, porcelain

number, kind and variety of such materials are extensive but can generally be classified into organic and inorganic.

All these classes have one characteristic in common, namely, they are all good insulators. Their electrical properties may not differ sufficiently to make them serve as the basic factor of selection. Governing parameters are likely to be physical, thermal or chemical. Other likely selection factors include cost and processability.

Organic bulk insulators

This class of materials is generally polymeric in nature and their behaviour is largely determined by their chemical composition and molecular structure. Table 16.2 gives a listing of the relevant properties of some of these materials. Like other polymers, these materials are characterised by their relatively low density and high flexibility and ductility. Because of their relatively low melting point, the maximum operating temperature of polymeric materials is limited to less than 500 K. For higher service temperatures other classes of materials have to be used, e.g. ceramics and mica.

Inorganic insulating materials

Inorganic insulating materials include materials such as asbestos, ceramics, glasses and micas. These materials appear in a variety of forms ranging from bulk insulators to fibrous cloths and papers. Organic binders may be used to form composites. The properties given in Table 16.2 are those of bulk materials. Although asbestos is flexible, ceramics and glasses are not, and have to be preformed before firing to the rigid state. Unlike polymers, ceramics can only be machined with some difficulty. Generally, glazing of ceramic insulators improves their mechanical strength and prevents moisture pickup. Due to their higher melting points, the maximum service temperature of ceramic and glass insulators is higher than that of polymers.

16.5 ALLOCATION OF WEIGHTING FACTORS

Previous discussions have shown that the main requirements for bulk insulating materials can be divided into upper limit, lower limit and target value properties, as defined in Chapter 13.

1. Dielectric constant—upper limit property requirement for applications like high voltage cables and lower limit property requirement for applications like capacitors.

TABLE 16.2

PROPERTIES OF SOME ELECTRICAL INSULATING MATERIALS

Material	*Dielectric constant (60 Hz)*	*Dielectric strength (volt/mil)*	*Volume resistivity (ohm/cm)*	*Dissipation factor (60 Hz)*	*Maximum operating temperature (K)*	*Tensile strength (MN/m^2)*	*Elongation (%)*	*Specific gravity*	*Thermal expansion coefficient m/m/K × 10^{-5}*	*Relative cost**
Organic materials										
Melamine	8·75	255	$1·5 \times 10^{12}$	0·06	488	49	0·9	1·48	3·5	1·0
Urea	8·2	260	$10^{12·5}$	0·04	408	55	1·0	1·5	5·2	1·2
Glass filled DAP	4·5	380	$10^{14·5}$	0·03	423	62	10	1·75		2·3
PTFE	2·1	480	10^{18}	< 0·0002	394	24	300	2·17	9·5	8·8
CTFE	2·7	550	$1·2 \times 10^{18}$	0·0012	388	49	200	1·8	14·4	16·8
ETFE	2·6	2000	10^{16}	0·0006	377	46	250	1·7	9	16·2
Polyphenylene oxide	2·6	525	10^{17}	0·0006	394	63	55	1·1	6·5	4·0
PVC	5	600	$10^{12·5}$	0·12	347	17	250	1·35	13	1·1
Polysulphone	3·1	425	10^{14}	0·001	454	70	75	1·24	5·6	5·4

Polyethylene (low density)	2·25	750	$> 10^{16}$	0·0005	316	14	400	0·92	20	0·86
Polypropylene	2·2	550	$> 10^{16}$	0·0005	378	35	150	0·91	8·6	0·7
Inorganic materials										
Asbestos	6	6	2×10^{13}	0·1	588	1910		2·4	1	
Porcelain	6·8	150	5×10^{13}	0·014	643	42		2·5	0·6	
Alumina ($Al_2O_3+SiO_2$)	9·0	350	10^{16}	0·001	1343	175		3·8	0·8	
Zircon ($ZrO_2 \cdot SiO_2$)	9·0	350	10^{15}	0·014	1143	91		3·6	0·5	
Boron nitride	4·2	1000	10^{14}	0·001	1073	25			0·43	
Fused silica	3·7	10	10^{13}	0·0002	1173	56		2·2	2·8	
Soda lime glass	7·5	0·4	10^{6}	0·04	523	14		2·5	45	
Muscovite mica	6·5	5000	10^{16}	0·0001	773	70		3·0	3200	

*Relative cost is cost relative to melamine which is considered as unity. Prices based on 1977 prices in USA.

2. Dielectric strength—lower limit property requirement.
3. Resistivity—lower limit property requirement.
4. Dissipation factor—upper limit property requirement.
5. Maximum operating temperature—lower limit property requirement.
6. Tensile strength—lower limit property requirement.
7. Elongation percent—lower limit property requirement.
8. Impact resistance—lower limit property requirement.
9. Thermal expansion—usually target value requirement for maximum compatibility with other components in the electrical apparatus. Lower values of thermal expansion are desirable in ceramics and glass to improve their thermal shock resistance.
10. Specific gravity—upper limit property requirement.
11. Cost—upper limit property requirement.

Insulating material selection will be carried out for two types of service as follows:

Case One. Selection of an insulating material for an aerospace system. Under these conditions, the insulating material has to endure the severe service conditions of high temperatures caused by aerodynamic heating or radiation from the engine. The operating temperature is expected to reach 450 K. Besides the high temperature requirements, the material must also endure mechanical stresses of at least 42 MN/m^2 and should be light in weight with a maximum allowable specific gravity of 3. Compatibility requirements specify a target value of $2 \cdot 3 \times 10^{-5}$ K for thermal expansion coefficient. Electrical design requirements are: dielectric constant less than 10 at 60 Hz, minimum dielectric strength 6 volts/mil, minimum volume resistance 10^{12} ohm/cm, and dissipation factor 0·1 max. at 60 Hz.

Case Two. Selection of an insulation material for computer cables. The conducting material is flexible flat aluminium cable which was selected for space saving, light weight and adaptability to special configurations and stresses. During installation, the cable will be folded through an angle of 180°. Service temperature will not exceed 350 K. Cost is an important consideration because large quantities of these cables will have to be used per machine. Due to space limitations, the dielectric strength should be greater than 100 and volume resistivity should be greater than 10^{12}. The dielectric constant should be less than 3·5 and the dissipation factor should be less than 0·01 to avoid excessive heating when the cables are stacked close together. Weight limitations restrict the specific gravity of the insulating material to less than 2·5. Compatibility requirements specify a target value for the thermal expansion coefficient of $2 \cdot 3 \times 10^{-5}$ K. The

TABLE 16.3

WEIGHTING FACTORS FOR ELECTRICAL MATERIAL PROPERTIES

Property	*Case One: Aerospace Application*	*Case Two: Computer Cable*
Dielectric constant	0·12	0·08
Dielectric strength	0·15	0·13
Volume resistivity	0·15	0·13
Dissipation factor	0·10	0·10
Maximum operating temperature	0·16	0·07
Tensile strength	0·08	0·07
Elongation percent	—	0·11
Specific gravity	0·16	0·10
Thermal expansion coefficient	0·08	0·08
Cost	—	0·13

elongation percent should be greater than 50% for the material to withstand the bending of the cable.

The weighting factors of the different material requirements were determined for the two types of service discussed here. The digital logic approach, described in Chapter 8, was used and the results are shown in Table 16.3.

16.6 EVALUATION OF INSULATING MATERIALS

Decision on the optimum material for insulation for the two types of service conditions under consideration was reached using the limits on properties method, as described in Chapter 13. In calculating the relative merit of each material, the log values of dielectric strength, volume resistivity, and dissipation factor were used.

Case One. Materials that did not meet any of the design requirements were eliminated from the list of candidate materials given in Table 16.2. For example, materials with maximum operating temperatures less than 450 K, tensile strength less than 42 MN/m^2, and specific gravity greater than 3 were eliminated. The electrical requirements were met by most materials in Table 16.2.

The materials that satisfied the design requirements were then evaluated and the results are shown in Table 16.4. Fused silica is the optimum

TABLE 16.4

EVALUATION AND GRADING OF ELECTRICAL INSULATING MATERIALS

Material	*Case One: Aerospace Application*		*Case Two: Computer Cable*	
	δ	*Preference*	δ	*Preference*
Organic materials				
Melamine	0·725	4	rejected	
Urea	rejected		rejected	
Glass filled DAP	rejected		rejected	
PTFE	rejected		0·829	5
CTFE	rejected		1·013	6
ETFE	rejected		0·80	4
Polyphenylene oxide	rejected		0·703	2
PVC	rejected		rejected	
Polysulphone	0·625	2	0·678	1
Polyethylene	rejected		rejected	
Polypropylene	rejected		0·712	3
Inorganic materials				
Asbestos	0·791	5	rejected	
Porcelain	0·701	3	rejected	
Alumina	rejected		rejected	
Zircon	rejected		rejected	
Boron nitride	rejected		rejected	
Fused silica	0·583	1	rejected	
Soda lime glass	rejected		rejected	
Muscovite mica	rejected		rejected	

material followed by polysulphone, then porcelain and melamine. The difficulty of processing silica may limit its use and if complex insulation shapes are required, the organic insulators will be preferable.

Case Two. As flexibility of the insulating material is an important requirement in this case, the ceramics and glass insulators were eliminated. Melamine, urea and glass filled DAP were also eliminated because of their lack of ductility. Asbestos was rejected because of its low dielectric strength. PVC was rejected because of its high dissipation factor and polyethylene because of its low maximum service temperature.

The materials that satisfied the design requirements were then evaluated and the results are shown in Table 16.4. Polysulphone is the optimum material followed by polyphenylene oxide and then polypropylene. Although polysulphone is relatively expensive, it has excellent electrical properties, high maximum operating temperature and the nearest expansion coefficient to the aluminium conductor.

BIBLIOGRAPHY

1. G.L. Moses, *Electrical Insulation,* McGraw-Hill (New York), 1951.
2. H.L. Saums and W.W. Pendleton, *Material for Electrical Insulating and Dielectric Functions,* Hayden (New York), 1973.
3. J.F. Young, *Materials and Processes,* 2nd Edn, John Wiley and Sons (New York), 1954.

17

Selection of Materials for Surgical Implants

17.1 INTRODUCTION

Surgical implant materials are used in repair of all anatomical systems in the body. Replacement of large portions of bones and a few other organs with metallic, ceramic, polymeric or composite prostheses is common. Soft tissues can also be restored or replaced with implants of similar texture and colour using polymeric materials. Mechanically, implants are relatively simple; however, biocompatibility requirements are difficult to meet. Bicompatibility means that the material and its possible degradation products must be tolerated by the biological tissue and cause no tissue disfunction at any time. It has been found that some people are significantly sensitive to given materials while others are not sensitive. There is a need to test patients for sensitivity to a particular material before implantation, and to retest them if implants have to be removed and replaced since sensitivity can be acquired in time. There are a large number of materials available for most implants and some may be more suitable than others in this regard for a given individual. For the present study this latter factor cannot be considered and selection will be made among materials that are generally accepted for surgical implants.

The possible positions and functions of the implant component inside the human body can be so different that material selection will be meaningless unless it is made for a specific job. For example, foams and gels are used for soft tissue supplementation; elastic materials for replacement of muscles, skin, cartilage, internal organs or blood vessels; rigid and strong materials for fixing or replacing bones and joints. This chapter will be devoted to the selection of surgical implant materials for hip joints. This is because the head of the femur is more frequently replaced than any other organ or structure in the body except teeth. The cause of this frequency is the very high incidence of fracture and arthritis of the hip in the ageing population. A typical prothesis is shown in Fig. 17.1.

17.2 DESIGN CONSIDERATIONS

Implants for hip joints are generally intended to be permanent and are expected to serve for years or decades. Bone, which the joint replaces, is a complex multiphase composite material of highly anisotropic properties.

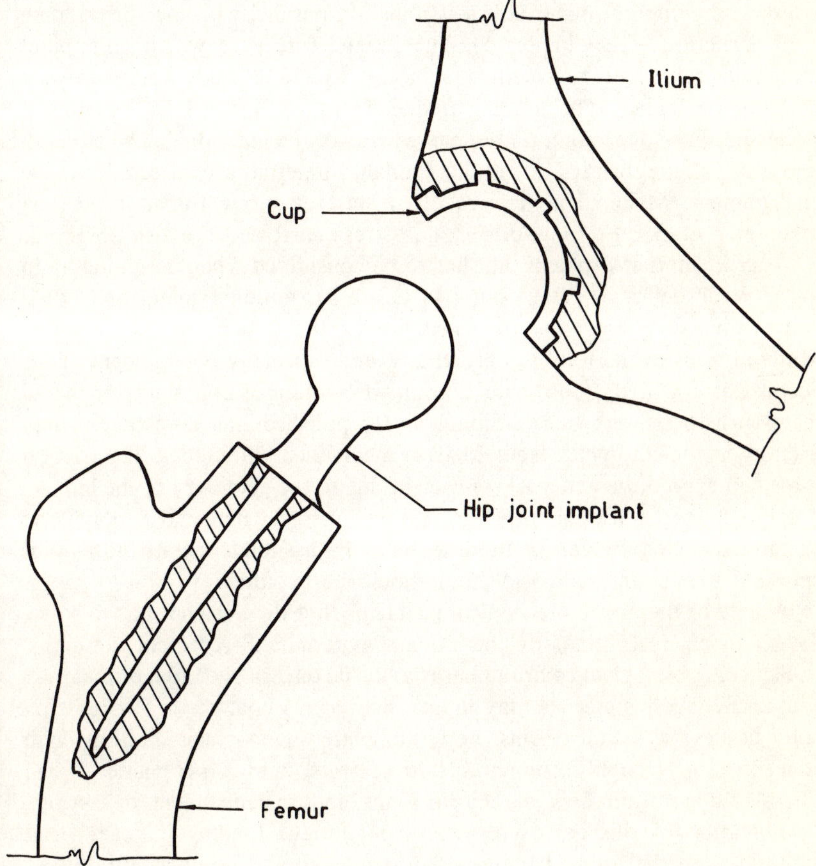

FIG. 17.1. Arrangement of implant material in the hip joint.

The compressive strength of bone is about 140 MN/m^2 and the elastic modulus is about 14 GN/m^2 longitudinally and about one-third of that radially. Bone is highly strain-rate sensitive and can fail on sudden loading. These properties are modest in comparison with those of most engineering metallic metals and alloys. However, live healthy bone is self repairing and has a great resistance to fatigue or repeated loading. The metal implant, on the other hand, does not have this ability to repair itself, and has a finite fatigue life. For this reason, the implant material must be much stronger than bone especially under fatigue loading. The situation is further aggravated by the fact that materials normally used in hip joints have a higher modulus of elasticity than bone, and this concentrates most of the stress in the implant material if it is joined in parallel with the surrounding bone structure (see Chapter 5). The fatigue forces to which an implant material is subjected are usually equal to 2·5 to 3 times body weight and can possibly be higher. The loading frequency ranges from 1 to 2·5 million cycles per year depending on the movements of the individual. The strength and the fatigue behaviour of the implant material are affected by the environment. Mechanical failures often start at screw thread roots and machining marks. Fretting, crevice and stress corrosion are also observed.

Wear is another problem that has to be considered when designing a hip joint. Wear debris is often found in tissue surrounding joints and could cause adverse effects due to sensitivity of the patient to the material. Material wear in the total hip prosthesis, and especially enlargement of the acetabular concavity (cup) causes poor articulation of the joint. For the all metal prosthesis, the ball, although finely polished has a relatively high friction coefficient, which leads to great mechanical difficulties. The friction force can exceed the patients own weight due to the geometry of the hip. As a solution to the friction problem, a small metallic ball and a polymeric acetabular concavity can be used. Combining dissimilar metals in the joint can lead to glavanic corrosion and should be avoided.

In spite of the above mechanical problems that the designer has to solve, the biological restrictions on the implant material still represent the major challenge. A biological environment is a hostile one for the implant material, conversely, such materials may induce deleterious changes in surrounding body tissues or distant organs. Body fluids are aqueous salt solutions with concentrations roughly comparable to seawater and a pH value of 7.4. There is a constant flow of chloride ions and replenishment of oxygen. Combination of this corrosive action and fatigue loading imposes strict limitations on the implant material surface finish and homogeneity. Stress corrosion cracking of metallic implants is a common cause of failure. The action of the implant material on the body can range from toxicity, in which case the implant material is totally rejected, to inertness. Even with inert implants, their presence impedes the normal healing sequence at the implant site and leads to a fibrocartilagenous membrane of low cellularity which isolates the implant from normal tissue. This membrane is the body's response to the stimulus of an inert foreign material which is impervious to body fiuid. The thickness of the membrane is proportional to the degree of metallic dissolution. Toxicity is usually signalled by a large population of inflammatory cells.

17.3 ANALYSIS OF IMPLANT MATERIAL REQUIREMENTS

From the above discussions it becomes clear that the hip joint implant material requirements are very stringent and only a few metals and alloys can meet these requirements. The main functional requirements for a hip joint implant are: adequate static and fatigue strengths, high hardness, elastic properties comparable to those of the surrounding bone, and weight and density comparable to that of the bone. The alternating stresses on the head of the femur when a person is walking can range from 2·5 to 3 times the body weight at a frequency of about 1 to 2·5 million cycles per year. The situation is aggravated by the presence of corrosive body fluids. Higher fatigue strength of the implant material means longer life and less probability of failure. High static strength gives the designer more freedom in selecting his cross sectional areas and his angles, and reduces the chance of plastic deformation or even fracture under unexpected loading conditions. Higher hardness of the implant material corresponds to higher wear resistance especially when metal to metal abrasion is involved. Elastic compatibility of implant material with the surrounding bone structure is important and large mismatches can lead to a serious deterioration of the interface between the two systems. Unfortunately, the elastic moduli of the currently available metallic implant materials are much higher than that of bone. This problem has to be solved by proper design and proper cementing agents between the bone and implant, but the solution will be easier for implants with Young's modulus nearer to that of the bone. Also, lower Young's modulus means higher damping capacity and resilience. The specific gravity of compact human bone is about 2·1 which is less than that of the implant metallic materials in current use. To keep the weight of the implant joint as near as possible to the original bone, high strength to weight ratio is important, but geometric requirements may necessitate a certain volume of the material in which case low density is important.

Processability requirements of hip joint implant materials are: soundness and fluidity for cast alloys, and sufficient ductility at forming temperatures for wrought alloys. A general requirement for all hip joint implants is homogeneity of structure and composition to avoid pitting or galvanic corrosion. Mirror finish for contact surfaces is also important.

The finished hip joint should be relatively inexpensive in order to make it available for use for the entire populace. Unless there is a special need which otherwise cannot be met, it would be logical to consider the 'cost on the job' of the material as one of the requirements.

Service requirements are of special importance in the case of implant materials as was discussed earlier. Biocompabitiblity, non-toxicity, chemical stability and inertness to the corrosive action of body fluids are essential if the material is to be considered for implant purposes. These requirements are difficult to quantify and materials will be given a rating of 10 for the best and 1 for the worst.

17.4 CLASSIFICATION OF METALLIC IMPLANT MATERIALS

Four principal metal systems can be considered for hip joint implants: stainless steels, cobalt–chromium-base alloys, titanium and titanium alloys and tantalum. These systems combine adequate mechanical properties and biocompatibility. Although some polymeric materials are extensively used in surgical implant applications, none of them are strong enough to bear the severe mechanical conditions acting on the hip joint. Some ceramics combine biocompatibility and mechanical strength, however their brittleness and lack of toughness represent an obstacle to their use in hip joint implants. The compositions of the four metallic systems are given in Table 17.1. Stainless steels depend on the presence of a closely adherent oxide surface layer for their resistance to corrosion. When continuous, this thin layer protects the metal from any further oxidation. Stainless steels represent the least inert implant metal system while titanium and its alloys are the most inert. Table 17.2 gives a grading of the corrosion resistance of the different materials. Titanium and its alloys also have the least elastic modulus and density in the list, which is an advantage, however their tissue tolerance is least. Cobalt-base alloys have excellent biocompatibility and strength but are less ductile than stainless steels.

TABLE 17.1

COMPOSITION OF METALS AND ALLOYS USED IN SURGICAL IMPLANTS (%)

No.	*Material*	*Cr*	*Ni*	*Co*	*Mo*	*Fe*	*C*	*Mn*	*Si*	*Ti*	*Others*
1	316 Stainless steel	17–20	10–14		2–4	Rem.*	0·03–0·08	2·0	0·75		
2	321 Stainless steel	17–19	9–12			Rem.*	0·08	2		0·4	
3	347 Stainless steel	17–19	9–13			Rem.*	0·08	2			0·8 Cb–Ta
4	317 Stainless steel	18–20	11–15		3–4		0·08	2	1·0max		
5	Cast Co–Cr alloy	27–30	2·5 max	Rem.*	5–7	0·75 max	0·36 max	1max	1max		
6	Wrought Co–Cr alloy HS–25	19–21	9–11	Rem.*	0·13	3max	0·05–0·15	2max	0·48		14·6–16·0 W
7	Unalloyed titanium					0·3	0·08			Rem.*	0·13 O_2, 0·07 N_2
8	Ti–6Al–4V alloy					0·25	0·08			Rem.*	5·5–6·5 Al, 3·5–4·5 V
9	Tantalum				0·01	0·01	0·01		0·005	0·01	0·01 N_2, 0·015 O_2

*Rem. = remainder

TABLE 17.2

PROPERTIES OF METALLIC IMPLANT MATERIALS

No.	*Material*	*Yield strength* (MN/m^2)	*Tensile strength* (MN/m^2)	*Elongation (%)*	*Modulus of elasticity* (GN/m^2)	*Rockwell hardness (RB)*	*Endurance limit 10^7 cycles in air* (MN/m^2)	*Inertness (corrosion resistance)*	*Tissue tolerance*	*Specific gravity*	*Relative cost**
1	316 Stainless steel (annealed)	207	517	40	200	80	350	7	10	8·0	1
1	316 Stainless steel (cold worked)	689	862	12	200	88	570	7	9	8·0	1
2	321 Stainless steel (annealed)	250	610	55	200	80	408	7	9	7·9	1
3	347 Stainless steel (annealed)	250	650	50	200	84	430	7	9	8·0	1
4	317 Stainless steel (annealed)	280	630	50	200	85	415	7	9	8·0	1
5	Cast Co–Cr alloy	450	655	8	238	100	438	9	10	8·3	3·1
6	Wrought Co–Cr alloy HS-25	379	896	55	242	99	600	9	10	9·13	3·1
7	Titanium	485	550	15	110	80	315	10	7	4·5	2·7
8	Ti–6Al–4V	830	985	10	124	83	490	10	7	4·43	2·8
9	Tantalum (cold worked)	345	480	15	190	75	300	9	8	16·6	75

*Relative cost is calculated relative to 316 stainless steel based on 1976 prices.

17.5 ALLOCATION OF WEIGHTING FACTORS

Previous discussions have shown that the main requirements for a hip joint surgical implant can be summarised as:

1. Yield strength.
2. Tensile strength.
3. Ductility–elongation percent.
4. Modulus of elasticity.
5. Hardness.
6. Endurance limit.
7. Corrosion resistance.
8. Tissue tolerance.
9. Specific gravity.
10. Cost.

As defined in Chapter 13, the yield strength, tensile strength, ductility, hardness, and endurance limit are all lower limit properties and it is better if the material has as high a value as possible. The modulus of elasticity and specific gravity should be as close as possible to those of bone, i.e. metals with lower values are better. The corrosion resistance and tissue tolerance represent the biocompatability parameters and should be as high as possible. The cost represents the total cost on the job of the implant material and includes the raw material as well as the processing costs. This should be as low as possible.

The relative importance of the above parameters is assessed using the digital logic approach described in Chapter 8 and the different weighting coefficients are given in Table 17.3.

17.6 EVALUATION OF IMPLANT MATERIALS

Decision on the optimum material for the hip joint implant material was reached using the weighted properties method, which was discussed in Chapter 13. All the materials in Table 17.1 are in use as implant materials, so that none of them will be rejected as unsuitable but will only be graded according to their merit for hip joint applications. Table 17.4 gives the results of evaluation and the order of preference of the different materials.

TABLE 17.3

WEIGHTING FACTORS FOR HIP JOINT IMPLANTS

Property	*Weighting factor*
Yield strength	0·08
Tensile strength	0·08
Ductility	0·08
Modulus of elasticity	0·08
Hardness	0·11
Endurance limit	0·15
Corrosion resistance	0·15
Tissue tolerance	0·15
Specific gravity	0·06
Cost	0·06

The material with the highest performance index (γ) is the Ti–6Al–4V followed by the Co–Cr wrought alloy and then the cold worked 316 stainless steel. The titanium alloy has good mechanical properties, excellent corrosion resistance and light weight. If the implant material is intended for temporary use, the distribution of weighting coefficients would put less emphasis on the corrosion resistance and fatigue strength and more emphasis on the cost, and this would change the balance in favour of stainless steels.

TABLE 17.4

EVALUATION AND GRADING OF IMPLANT MATERIALS

Material	γ*	*Preference*
Ti–6Al–4V (annealed)	81·2	1
Wrought Co–Cr alloy (HS-25)	79·3	2
316 Stainless steel (cold worked)	77·5	3
347 Stainless steel (annealed)	73·1	4
317 Stainless steel (annealed)	72·9	5
321 Stainless steel (annealed)	72·5	6
Titanium	71·6	7
Cast Co–Cr alloy	70·4	8
316 Stainless steel (annealed)	67·4	9
Tantalum (cold worked)	57·6	10

*See Section 13.3 for method of calculating γ.

BIBLIOGRAPHY

1. A.C. Fraker and A.W. Ruff, *J. Metals*, **29**(1977), 22.
2. C.A. Homsy, *J. Biomedical Materials*, **4**(1970), 341.
3. J.L. Katz, *Biomaterials Proceedings*, Seminar-Workshop in Biomaterials, Battell Seattle Research Center, A.L. Bement, Ed., University of Washington Press (Seattle), 1971, p. 79.
4. R.M. Rose, Orthopedic Implants, *Biomedical Physics and Biomaterials Science,* H.E. Stanley, Ed., MIT Press (Cambridge), 1972, p. 169.
5. G. Schmeisser, *J. Materials,* **3**(4) (1968), 951.

18

Selection of Materials for Gas Turbine Blades

18.1 INTRODUCTION

The gas turbine is a complex engineering product which may run for tens of thousands of hours before overhaul. The output may range from a few to many thousands of horsepower. The functions of a gas turbine vary from providing direct propulsion of conventional aircraft (jet), to propelling ships, trains and helicopters. It can be used to pump gas in transmission lines and to provide peaking and base electrical power in generating systems. Over the past 20 years, power output and specific fuel consumption have been improved markedly by materials changes and engineering design.

The major sections of a gas turbine are the compressor, combustion section, shaft and turbine. In advanced engines, the turbine may have as many as six stages; the first two are termed high pressure and the rest are low pressure. These stages convert the thermal energy supplied by the combustion into mechanical energy and must withstand the full impact of the hot gases coming out of the combustion chamber. The nearer the operating temperature of the turbine material to the gas temperature, which can exceed 1600 K, the higher the efficiency of the engine. The turbine operating temperature can be increased by 1, employing better heat resistant alloys, 2, internal cooling of the turbine parts and 3, use of coatings. Of all the turbine parts, the blades are subjected to the most severe conditions and the development of better materials for their manufacture has always been a challenge to the materials engineer. Although good resistance to creep deformation and rupture has often been the prime requirement, achievement of creep strength alone does not present the major problem. The greatest difficulty is avoidance of undue deterioration in other properties of the material as creep resistance is improved. Toughness, oxidation and corrosion resistance and processability usually suffer when alloying elements are added to improve creep strength. Lack of toughness has been one of the main reasons for not using ceramic materials in turbine blades so far. Reduced oxidation and corrosion resistance can be overcome by surface coatings but processability remains a major problem in view of the increasingly more complex shapes that have to be developed to improve the cooling of the blade.

18.2 DESIGN CONSIDERATIONS

The service conditions in which the turbine blades operate are severe. The gases that impinge on the blades are the products of combustion of air and kerosene, with an excess of air which makes them corrosive and oxidising. Fuel ashes containing sulphur and excess O_2 are the main active ingredients. The temperature of these gases as they enter the turbine can be in excess of 1600 K in some cases which means that the turbine blades are operating in a gas the temperature of which can be above the melting point of the material. At the same time the blades are subjected to stresses in the range of 100 to 160 MN/m^2. These severe conditions can only be met by the utilisation of air cooling. Although high stress rupture strength of the material is important to avoid failure in service, creep resistance is necessary to avoid the possibility of blade rubbing. Another complicating factor is that the blade material is also subjected to thermal shocks and thermal fatigue which accompany the starting and shutting down of the engine. The temperature changes in the blade material can be so rapid that appreciable differences are developed between the temperature of the surface regions, which heat or cool rapidly, and that of the inner core of the blade. This results in strains in the blade, and if the strain is locally greater than the ductility of the material, the blade will crack in that region (thermal shock), even if the strain imposed by a single change of the temperature is below the fracture strain, repeated cycling can build up strain damage and lead to eventual cracking (thermal fatigue).

18.3 MATERIAL REQUIREMENTS

The above discussions indicate that the service conditions in which a turbine blade operates are the most difficult in the gas engine. This explains why the turbine blade material represents the ultimate in gas engine materials. The prime functional requirement of turbine blade materials is creep behaviour which covers creep resistance, stress rupture strength and stress rupture ductility. Creep resistance can affect the service life of the blade as excessive creep strains can cause rubbing with stationary parts. However, stress rupture ductility is important in order to avoid failure when the material cannot accommodate even the allowable small amount of creep deformation. Also, creep rupture stress for the expected blade life should be more than the stresses encountered in service so as to avoid blade failure. Stability of the structure and properties can be one of the more difficult requirements to guarantee. Grain growth, coarsening of phases and spheroidisation are among the changes that can cause deterioration of properties after long exposure to high temperatures, as discussed in Chapter 6. Unfortunately, creep resistance and stability are often obtained only with some sacrifice of ductility so that only a compromise is usually possible.

Another functional requirement is thermal shock and thermal fatigue resistance. Besides the thermal strain imposed on the material at the starting and shutting down of the engine, additional strains develop due to air cooling of the blade. Although no simple relation has been found between thermal fatigue resistance and creep properties, an empirical relation has been proposed between thermal fatigue resistance and tensile properties of alloys:

$$\text{Number of cycles to cracking} = \frac{16 \times 10^8 \sigma_T \varepsilon}{(T - 600)^4} \qquad (18.1)$$

where σ_T = yield strength at maximum temperature of the cycle (T), ε = tensile elongation at temperature 100° C below (T).

Processability requirements of the material depend on the intended means of manufacture of the blade. Forging, extrusion, precision casting and directional solidification are the main processing methods for turbine blades. The presence of complex internal channels for air cooling represent a challenge to the production engineer. Wrought alloys should exhibit enough ductility at the working temperature, while cast alloys should be free from porosity and segregation. Highly specialised casting methods have been used to improve the creep properties of cast alloys, and solidification control has been improved to produce single crystal blades.

Cost used not to be an important parameter in the early tages of gas turbine development, but the competition now is such that even in aerospace applications, cost reduction is essential. In the case of turbine blades, the processing, surface coating and finishing costs form a sizeable fraction of the finished blade cost. In aerospace applications reliability is vital and inspection costs are important.

High corrosion and oxidation resistance are the major service requirements for blade materials. These requirements become vital in applications like air-cushion vehicles and trains, where economics dictate the use of impure fuels and the ingestion of contaminants such as seawater. Sacrifice of corrosion and oxidation resistance is usually part of the compromise necessary to achieve high creep strength. This problem has been partially overcome by surface coating the blade and many blades are now aluminised, i.e. given a coating of nickel aluminide to protect the surface from oxidation. However, at temperatures in excess of 1300 K interdiffusion rapidly increases and limits the use of this type of coating. Non-diffusion type coatings have the advantage that the oxidation resistant aluminium is not lost by interdiffusion to the substrate. These coatings can be applied by vacuum deposition and last more than $1\frac{1}{2}$ times as long as aluminide coatings.

18.4 CLASSIFICATION OF BLADE MATERIALS

Previous discussions have shown that turbine blades face the most demanding service conditions within the gas turbine because of the combination of extreme mechanical stresses and high temperature operation. The use of materials with high melting points is obviously desirable. Ceramics combine high melting points and oxidation resistance, but their extreme brittleness has so far outweighed their potential advantages. The refractory metals, Nb, Mo, W and Ta have high melting points but they are too susceptible to oxidation to be used as turbine blades. Even if suitable alloys of these metals are developed, cost would be a severe adverse factor. Present turbine blade materials are mostly either nickel-base or cobalt-base alloys with the nickel-base alloys being used for most applications.

Table 18.1 gives some of the alloys that are usually recommended for turbine blade applications. These alloys can be generally divided into wrought alloys and cast alloys. Generally, wrought alloys are more ductile but can be expensive to machine, and can become difficult to work as their compositions are developed to provide better high temperature strength. Cast blades offer more flexibility in design at the expense of ductility and with inherent problems of casting porosity and dross films. However, casting approximately to size can save a considerable amount of the machining cost, especially in complex compounds. With the larger additions

TABLE 18.1

CHEMICAL COMPOSITION OF SOME GAS TURBINE BLADE MATERIALS

Material	*Nominal composition, weight %*											
	Ni	*Cr*	*Co*	*Mo*	*W*	*Ta*	*Nb*	*Al*	*Ti*	*C*	*Zr*	*Fe*	*Others*
Cast alloys													
713C	Balance	13		4			2	6	1	0·1			
M22	Balance	6		2	11	3		6		0·1			
MAR–M246	Balance	9	10	2·5	10·0			5·5	1·5	0·15	0·05		0·1 Mn, 0·05 Si
MAR/M302		21·5	58·0		10·0	9·0				0·85	0·2		
MAR–M322		21·5	61·0		9·0	4·5			0·75	1·00	2·5		
MAR/M509	10	23·5	55		7·0	3·5			0·2	0·6	0·5		
NASA Co W Re		3	68		25				1·0	0·4	1·0		2·0 Re
WI–52		21	63		11	2				0·45		2	0·25 Mn, 0·25 Si
B1900	64	8·0	10	6·0		4·0		6·0	1·0	0·1	0·1		
Wrought alloys													
Nimonic 75	78·8	20·0							0·4	0·01			0·1 Mn, 0·7 Si
Nimonic 80A	74·7	19·5	1·1					1·3	2·5	0·06			0·1 Mn, 0·7 Si
Nimonic 90	57·4	19·5	18·0					1·4	2·4	0·07			0·5 Mn, 0·7 Si
Nimonic 105	53·3	14·5	20·0	5·0				1·2	4·5	0·2			0·5 Mn, 0·7 Si
Nimonic 115	57·3	15·0	15·0	3·5				5·0	4·0	0·15			
Rene 80	Balance	14	9·5	4·0	4·0			3·0	5·0	0·17	0·03		
Waspaloy A	58·3	19·5	13·5	4·3				1·4	3·0	0·07	0·09	2·0	0·5 Mn, 0·5 Si
Udimet 700	53·4	15	18·5	5·2				4·3	3·5	0·08			
T–D Nickel	98·0												2·0 ThO_2

of alloying elements, casting alloys generally exhibit higher stress rupture strengths than wrought alloys, as shown in Table 18.2. Sacrifice of oxidation resistance is part of the compromise necessary to achieve the high creep strengths offered by the cast alloys, as shown in Table 18.2. The reason for this sacrifice is that increasing Ti and Al in order to improve the creep strength causes the precipitation of the brittle sigma phase during exposure to high temperatures. The presence of the sigma phase is detrimental to stress-rupture and impact properties. This undesirable constituent can best be avoided by reducing the Cr content, but this reduces oxidation resistance.

The most important advances in the development of casting alloys have been vacuum melting and casting, and also directional solidification. In the latter technique all the grains are aligned longitudinally along the length of the blade which improves the stress rupture life. Directional solidification has been used to produce single crystals where the strengthening phase is aligned along the length of the blade. Table 18.1 does not include any of these alloys due to lack of information.

Dynamic burner-rig testing shows that nickel-base alloys are superior in oxidation resistance to cobalt-base alloys. In general, the higher the reactive element additions (Al, Si and rare earths) and the lower the refractory metal (Mo, W, Ta and Nb) content, the greater the oxidation resistance, regardless of the base. While nickel alloys owe their superior oxidation resistance to oxide scales which are generic to α-Al_2O_3, conventional cobalt-base alloys do not contain aluminium and depend upon α-Cr_2O_3 for protection. These principles were applied in grading the alloys of Table 18.2 in order of their oxidation resistance. This grading is very approximate as the figures in the literature were not obtained under the same test conditions. Alloys with the highest oxidation resistance are given a grade of 100.

A rough estimate of the relative cost on the job of blades made from different materials is given in Table 18.2. Although some wrought alloys may be less expensive to buy in the form of stock, the more expensive forming and machining involved may change the relative cost of the finished blade. The relative thermal fatigue resistance of the different materials is calculated according to Eqn. (18.1) and is given in Table 18.2.

18.5 ALLOCATION OF WEIGHTING FACTORS

Previous discussions have shown that the main requirements for turbine blade materials can be summarised as:

1. Specific tensile strength at operating temperature.

TABLE 18.2

PROPERTIES OF SOME GAS TURBINE BLADE MATERIALS

Material	*Specific gravity*	*UTS at 871°C (MN/m^2)*	*Yield at 0·2% offset at 871°C (MN/m^2)*	*Elongation % at 760°C*	*Rupture strength for 100h life at 1000°C (MN/m^2)*	*Oxidation resistance rating*	*Relative cost on the job of a blade**	*Relative thermal fatigue resistance*	*Specific UTS*	*Specific rupture strength*
Cast alloys										
713C	7·91	725	497	6	147	60	55	21	91·7	18·6
M22	8·63	884	676	5	172·7	10	66	23	102·4	19·9
MAR–M246	8·44	862	690	5	189	35	51	24	102·1	22·4
MAR–M302	9·21	448	310	8	112	20	100	17	48·6	12·2
MAR–M322	8·91	552	345	6·5	140	40	78	15	62·0	15·7
MAR–M509	8·85	352	290	10	119	50	70	20	39·8	13·4
NASA Co W Re	9·59	496	350	2	126	10	58	5	51·7	13·1
WI–52	8·88	414	276	9	80·5	40	65	17	46·6	9·1
B1900	8·22	794	696	4	182	35	74	19	96·6	22·1
Wrought alloys										
Nimonic 75	8·37	145	110	70	9·42	100	65	53	17·3	1·1
Nimonic 80A	8·22	400	262	20	30	100	67	36	48·7	3·6
Nimonic 90	8·18	428	262	10	40	100	69	18	52·3	4·9
Nimonic 105	7·99	607	366	22	56·5	70	75	55	76·0	7·1
Nimonic 115	7·85	828	552	22	100·5	75	80	83	105·5	12·8
Rene 80	8·24	621	552	11	95	60	85	41	75·4	11·5
Waspaloy A	8·19	525	518	28	47·6	80	78	100	64·1	5·8
Udimet 700	7·91	690	635	20	119	75	79	88	87·2	15·0
TD Nickel	8·9	193	179	11	98	85	76	14	21·7	11·0

*Relative cost is calculated in relation to the cost of MAR–M302 which is taken as 100. The prices are based on 1977 prices in

2. Specific rupture strength at operating temperature.
3. Oxidation resistance at operating temperature.
4. Cost.
5. Thermal fatigue.

The use of specific tensile strength and specific rupture strength (strength/specific gravity) is important when selecting turbine blades as they usually rotate at very high speeds, which could cause excessive centrifugal forces if the material is heavy. Also, if the gas turbine is to be used in aircraft applications, weight saving becomes important. The strength properties represent the most important parameters in selection. Although oxidation resistance is an important requirement, the use of surface coatings for protection can reduce its weighting factor.

Cost of the material alone is not the main parameter to consider in the case of turbine blades. Processing costs represent a major portion of the total cost of the finished blade, especially for complex blade shapes where internal air cooling is employed.

Thermal fatigue is an important parameter if blade cracking in service is to be avoided; it is also important because it involves the yield point and elongation percent of the material and these are important design parameters in their own right. The relative importance of the above parameters was determined using the digital logic approach, which was described in Chapter 8, and the results are shown in Table 18.3.

TABLE 18.3

WEIGHTING FACTORS FOR GAS TURBINE BLADE MATERIALS

Property	*Specific strength*	*Specific rupture strength*	*Oxidation resistance*	*Thermal fatigue resistance*	*Cost*
Weighting factor	0·2	0·27	0·17	0·23	0·13

18.6 EVALUATION OF GAS TURBINE BLADE MATERIALS

Evaluation of the materials listed in Table 18.1 was carried out using the weighted properties method, which was described in Chapter 13. The cost was considered as one of the properties. The material performance index (γ) of the different alloys is given in Table 18.4, together with the order of

TABLE 18.4

EVALUATION AND GRADING OF GAS TURBINE BLADE MATERIALS

Material	γ	*Preference*
Cast alloys		
713C	67·0	4
M22	60·4	7
MAR-M246	70·9	3
MAR-M302	40·0	17
MAR-M322	49·6	10
MAR-M509	46·4	13
NASA Co W Re	39·9	18
WI–52	40·6	16
B1900	64·0	5
Wrought alloys		
Nimonic 75	43·8	15
Nimonic 80A	48·7	11
Nimonic 90	46·5	12
Nimonic 105	56·4	8
Nimonic 115	75·6	2
Rene 80	55·9	9
Waspaloy A	64·3	6
Udimet 700	76·0	1
TD Nickel	44·2	14

preference which is based on the value of γ. Udimet 700 was found to be the optimum material followed by Nimonic 115 then MAR-M246.

It is to be noted that the list of materials used for selection did not include any of the directionally solidified alloys. These materials are relatively new and many of them are still in the development stages. However, directionally solidified materials are expected to be strong competitors to the alloys discussed in the present case study.

BIBLIOGRAPHY

1. E.F. BRADLEY, D.G. PHINNEY and M.J. DONACHIE, *Metal Progress,* **98**(1970), 68.
2. J.H. CAIRNS and P.T. GILBERT, *The Technology of Heavy Non-ferrous Metals and Alloys*, George Newnes Ltd (London), 1967.
3. F.J. CLAUSS, *Engineering Guide to High Temperature Materials*, Addison-Wesley (Reading Mass., London, Amsterdam), 1969.

4. M.J. Donachie and E.F. Bradley, *Metal Progress*, **96**(1969), 60.
5. M.J. Fleetwood, *Newer Engineering Materials*, R.F. Winters, Ed., Macmillan (New York), 1969.
6. H.E. Gresham, *Metals and Materials,* **3**(1969), 433.
7. *Metal Progress*, *Data Book*, American Soc. for Metals, 1977.
8. C.T. Sims and W.C. Hagel, *The Superalloys*, John Wiley and Sons (New York), 1972.

19

Selection of Materials for Low Temperature Applications

19.1 INTRODUCTION

In some industrial applications, low temperatures are used when a liquid may be preferable to a gas, e.g. transportation of liquid gases, or when low reaction velocity may be preferable to high reaction velocity, e.g. preservation of biological materials. Also the high molecular order associated with low temperatures is of importance both for basic research and technological applications, e.g. superconduction associated with the complete disappearance of electrical resistance in certain metals and alloys.

Low temperature applications can generally be divided into two broad categories, refrigerating and cryogenic. It is difficult to assign a definite temperature for the dividing point between refrigerating and cryogenic applications, but it will probably conform to present usage to consider cryogenic applications to be below 123 K (–150 °C).

When selecting a material for use in a piece of cryogenic equipment, a large number of factors must be considered and the final choice often involves a compromise between conflicting requirements, e.g. the material should be strong, have a high or low thermal or electrical conductivity, be weldable, and be economical. The prime characteristic of most equipment designed for use at low temperatures is that it should be tough, i.e. that it should not be liable to failure in a brittle manner. This is because of the ductile–brittle exhibited by some steel and several plastics at low temperatures.

19.2 DESIGN CONSIDERATIONS IN LOW TEMPERATURE SYSTEMS

Besides the usual design considerations of function, materials, manufacturing and cost, low temperature systems must meet strict safety requirements. There are two general areas in which hazards can occur. One of these areas is related to the properties of cryogenic fluids that are usually handled in the system, while the other area is related to the changes in the properties of materials of system components subjected to the low temperature environment.

In storage of cryogenic fluids, it must be remembered that they only exist as liquids at temperatures considerably below ambient. Suitable pressure relief systems should be provided to avoid explosions that can result from pressure build up due to vaporisation, caused by heat input from the environment. Uncontrolled outleakage of cryogenic fluids is also hazardous as they can be asphyxiating, e.g. N_2 and He, toxic, e.g. F and CO, or explosive, e.g. H_2 and CO. Direct physical contact with either the cryogenic liquid or cold vapour as well as with cold metal surfaces can cause serious tissue damage. These hazards can be overcome by taking the necessary design and operation precautions.

Low temperatures can affect the properties of the solid materials used in cryogenic systems. The strength of materials is not usually a problem as it increases with decreasing temperature. However, it is advisable to use room temperature strength in designing low temperature components. This is because the equipment may warm up while under stress and because temperature gradients may exist. Consideration must be given to the possible loss of ductility at low temperatures, also hydrogen gas can cause embrittlement at ambient temperatures.

During cooling down of low temperature systems, internal stresses can arise due to thermal gradients. Higher coefficients of thermal expansion, higher elastic moduli and lower thermal conductivities of the material lead to higher internal stresses. Also, lower expansion coefficients will reduce equipment displacement on cooling down to operating temperatures. Decreasing the temperature of solid materials generally lowers their specific heat which reduces their effectiveness in retarding a rapid temperature rise for a given heat input, however, the cryogen is the main heat sink in the storage system.

In general, hazards are expected to increase as the size of the cryogenic installation increases which makes the safety precautions very important in storage and transfer systems. The following discussions will be related to this type of cryogenic system. For cases where contamination of the cryogen is hazardous, it is necessary to remove undesirable gases from the storage vessel and to prevent inward leakage by maintaining the system above atmospheric pressure. As an additional precaution in long term storage of liquid cryogens, periodic warm-up and analysis of the vessel contents is desirable. The warm-up should bring the coldest spot of the vessel interior to a temperature higher than the boiling point of the suspected impurity. This is necessary to prevent excessive accumulation of solid air which there is no way of detecting or removing under normal cryogen storage conditions. Automatic pressure relief must also be provided in strategic locations in the vessel. Common practice is to use a safety relief valve set at approximately 1·1 times the working pressure, and a safety head or rupture device that will burst at about 1·2 times the working pressure.

Vessels used for storing and transporting cryogenic fluids can be classified according to the type of insulation: 1, vacuum insulation; 2, evacuated-powder and evacuated-fibre insulation; 3, rigid-cellular or fibre insulation. Types 1 and 2 have two or more concentric metallic shells, the innermost shell being the fluid container and the outer shell being the vacuum jacket. To ensure operating safety of the vessel, a maximum working stress of about 25% of the minimum tensile strength of the vessel material is allowed, (Section VIII of the ASME Code), or two-thirds of the yield stress (BS 1515). The wall thickness of the liquid shell of a vessel is a function of the pressure rating which should be as low as practicable to reduce the cost of the shell and the cool-down losses. While a stationary cryogenic vessel is only expected to bear the weight of the liquid, a transportation vessel has to bear the added forces of transportation, usually acceleration forces. For example, if the acceleration forces are three times the acceleration of gravity, the vessel should be designed to withstand three times the normal weight of the unit.

The heat leak to a cryogenic vessel is the quantity of heat per unit time that penetrates the container causing evaporation of the cryogenic liquid. This heat leak is a measure of the thermal efficiency of the vessel and is the summation of the leak through the insulation, the supports of the inner shell, the interconnecting piping and any accesses or manways. Selection of materials with low thermal conductivity is an important factor in reducing the heat leak.

The present case study will be concerned with the selection of the shell material for a storage tank that will operate at 77 K (the boiling point of N_2). The tank is portable so that light weight is important and, to avoid accidental breakage in moving and transporting, the material should be tough.

19.3 MATERIAL REQUIREMENTS

Although material requirements for most cryogenic applications are expected to be similar, the following discussions will be primarily concerned with storage and transfer vessels. Retention of ductility at the operating temperature is of prime importance if the material is to be used at cryogenic temperatures. Carbon steels are notorious for their low temperature brittleness, but most other structural metals such as aluminium, copper, nickel and their alloys as well as the series 300 austenitic stainless steels do not exhibit this phenomenon. It can be generally stated that the face-centred cubic metals and alloys do not become brittle at low temperature, while body-centred cubic metals do. Although the liquid shell of a cryogenic vessel may never be subjected to impact loading, stress concentration at minor local imperfections or stress raisers could lead to failure if not relieved by plastic deformation. Only a few polymers are known to be somewhat ductile at cryogenic temperatures, e.g. polytetrafluoroethylene (teflon) but some glass fibre reinforced polymers show good, low temperature properties. Clearly other mechanical properties like ultimate tensile strength, yield strength, hardness and fatigue resistance are important requirements but they present no particular problem as they all increase with decreasing temperature, which means that, from the strength point of view, a vessel that is strong enough at ambient temperature will be safe at cryogenic temperatures.

Although a lower elastic modulus is desirable in order to reduce thermal stresses, structural rigidity of a vessel requires a high value of elastic modulus. The danger of thermal stresses can be reduced by selecting a material with a relatively high thermal conductivity and a low thermal expansion coefficient. Slower cooling of the system can also reduce thermal stresses. The modulus of elasticity increases with decreasing temperature.

The decrease in the specific heat with decreasing temperature of most metals is not linear and the total cool-down loss below 70 K is only a small fraction of the losses from ambient temperatures to 70 K. Materials with lower specific heats cause lower cool-down losses. In comparing materials, it is more relevant to use the average specific heat per unit volume than the average specific heat per unit weight. The variation of the thermal expansion coefficient with temperature follows closely the variation of specific heat (Grüneissen relation) and most metals contract very little below 70 K, the bulk of contraction occurring on cooling from ambient temperature to 70 K. Materials with lower thermal expansion coefficients cause less internal stresses due to thermal gradients, but care should also be taken to avoid differential contraction between different components. Differential contraction can be severe if one component is made from certain non-crystalline solids, especially polymers. Anisotropy of contraction of metals with preferred orientation, polymers with long chain molecules, and fibre reinforced composites can lead to distortions on cooling and should be avoided.

Materials with lower thermal conductivities conduct less heat to the stored cryogen and are thus more desirable. The thermal conductivity of alloys and polymers decreases with decreasing temperature but in perfectly pure metals it should increase indefinitely. However, in practice the scattering due to impurities and defects dominates at very low temperatures and the conductivity passes through a maximum.

For a given shell diameter the shell thickness (t) for a given material is related to the operating pressure by the equation:

$$t = \frac{PD}{2se} \tag{19.1}$$

The cost of materials used for the construction of a storage shell can be related to the operating pressure and material parameters and is given by Haselden (see bibliography at end of chapter) by the equation:

$$TC = \frac{\pi CLPD}{se} \tag{19.2}$$

where TC = material cost of the shell, C = material cost per unit volume, L and D = shell length and diameter respectively, P = design pressure, e = joint efficiency, and s = working stress of the material.

For given shell dimensions and design pressure, the cost of two materials 1 and 2 can be compared as follows:

$$\frac{TC_1}{TC_2} = \frac{C_1 \cdot s_2 \cdot e_2}{C_2 \cdot s_1 \cdot e_1} \tag{19.3}$$

Besides the cost and working stress of the material, the joint efficiency is an important parameter in determining the total cost of a storage shell. Shells are usually produced from welded sheet metal. Flanged and bolted joints should be minimised to avoid leakage problems. Weldability of the shell material is thus an important parameter in the selection process. None of the commonly used metals and alloys present any real problems in welding, and gas shielded arc welding is usually employed.

19.4 CLASSIFICATION OF LOW TEMPERATURE MATERIALS

Face-centred cubic metals and alloys are the most common materials of construction for cryogenic applications. This is because they maintain their ductility at low temperatures. However, some of the titanium alloys, close packed hexagonal structures, remain moderately ductile at low temperatures and are useful for many applications. Even less ductile materials such as some body-centred cubic steels can be used under certain circumstances. Also, polymeric materials and composites which maintain some ductility at low temperatures can be used successfully. Some glass compositions are used in different cryogenic applications but will not be included in the present discussion as they are not practical in manufacturing large storage vessels. Table 19.1 gives a list of some materials that are commonly used in cryogenic applications and could be used as storage vessel shells. Table 19.2 gives a list of some relevant properties at room temperature and at 77 K, which was chosen as the working temperature for the storage vessel. The low temperature toughness data in the literature are not consistant and are often non-existent and to overcome the difficulty of comparing the present materials a toughness index (*TI*) was calculated as:

$$TI = \frac{UTS + Y}{2} \times \varepsilon \qquad (19.4)$$

where *UTS* is ultimate tensile strength of the material at 77 K, *Y* is yield at 77 K, and ε is elongation at 77 K.

Aluminium and its alloys are widely used in the construction of cryogenic systems as they combine high strength/weight ratios, ease of fabrication and low cost. They may be divided into two groups on the basis of their principal strengthening mechanisms, namely, solid solution hardenable (5000 series) and precipitation hardenable or heat treatable (2000 and 7000 series). In the optimum hardened condition the latter alloys are stronger but less ductile than the solid solution hardened alloys.

Copper and its alloys were among the first metals to be used in the construction of cryogenic equipment and they maintain excellent ductility at low temperatures. The main disadvantage of pure copper is its low yield strength and it is better to use copper alloys instead, e.g. α-brasses.

Nickel-base alloys have excellent corrosion resistance and toughness at cryogenic temperatures. Stainless steels also combine corrosion resistance, strength and weldability but sensitisation of welded areas can seriously im-

TABLE 19.1

COMPOSITION OF SOME ALLOYS USED IN LOW TEMPERATURE APPLICATIONS (%)

Material	*Al*	*Cu*	*Fe*	*Cr*	*Ni*	*Zn*	*Ti*	*Si*	*Mn*	*Mg*	*C*	*Others*
Aluminium alloys												
2014–T6	Rem.*	3·9–5·0	1·0 max	0·1 max		0·15 max		0·5–1·2	0·4–1·2	0·2–0·8		
5052–0	Rem.*	0·1 max	0·4 max	0·15–0·35				0·45 max	0·1 max	2·2–2·8		
5083–0	Rem.*	0·1 max	0·4 max	0·05–0·25			0·15 max	0·4 max	0·3–1·0	4–4·9		
5456–0	Rem.*	0·1 max		0·05–0·2			0·2 max	0·4 max	0·5–1·0	4·7–5·5		
7075–T6	Rem.*	1·2–2·0	0·7 max	0·18–0·4		5·1–6·1	0·2 max	0·2 max	0·3 max	2·1–2·9		
Stainless steels												
301 (full hard)			Rem.*	17	7							
304 (annealed)			Rem.*	19	10							
310–75% (cold worked)			Rem.*	25	20							
321 (annealed)			Rem.*	18	10		0·5					

410 (annealed)			Rem.*	12							
Nitronic 33 60% (cold worked)				18	3·25				13	0·05	0·3 N
Titanium alloys											
Ti–5Al–2·5Sn	4–6		0·5 max				Rem.*			0·1 max	2–3 Sn, 0·07 max N
Ti–6Al–4V	5·5–6·75		0·4 max				Rem.*			0·1 max	3·5–4·5 V, 0·07 max N
Superalloys											
Inconel X–750	0·7	0·2	7·0	15·5	Rem.*		2·5	0·25	0·5	0·04	1·0Nb
Inconel 718	0·5	0·2	18·5	19·0	Rem.*		0·9	0·2	0·2	0·04	3·0Mo, 5·0Nb
Copper alloys											
OFHC		99·95									
70Cu–30Zn		68·5–71·5	0·05 max			Rem.*					0·07 max Pb
70–30 Cupro-nickel (annealed)			0·4–0·7		29–33	1·0 max			1·0 max		0·05 max Pb

*Rem. = remainder.

TABLE 19.2

PROPERTIES OF SOME MATERIALS USED IN LOW TEMPERATURE APPLICATIONS

	*UTS at RT** *(MN/m²)*	*Yield at RT (MN/m²)*	*Elongation at 77 K (%)*	*Modulus of elasticity at RT (GN/m²)*	*Specific gravity*	*Thermal expansion average (10⁻⁶× K⁻¹)*	*Thermal conductivity at RT***	*Specific heat average between RT and 77 K***	*UTS at 77K (MN/m²)*	*Yield at 77K (MN/m²)*	*Cost $/lb (1971)*	*Toughness index†*
Aluminium alloys												
2014–T6	490	420	14	74·2	2·8	21·4	0·37	0·164	568	490	0·44	75·5
5052–0	196	91	46	70	2·68	22·1	0·33	0·164	308	105	0·46	95
5083–0	294	147	34	72	2·66	23·4	0·28	0·164	413	161	0·46	97·6
5456–0	315	161	32	72	2·66	22·1	0·28	0·164	441	203	0·48	103
7075–T6	581	511	6	74	2·8	21·8	0·29	0·164	707	623	0·5	40
Stainless steels												
301 (full hard)	1 575	1 365	40	189	7·9	16·9	0·04	0·08	2 205	1 645	0·46–0·76	770
304 (annealed)	525	210	20	189	7·84	17·0	0·04	0·08	1 540	420	0·46–0·76	196
310 (75% cold worked)	1 330	1 120	11	210	7·9	14·4	0·03	0·08	1 820	1 575	0·6	186·7
321 (annealed)	630	266	43	189	7·9	16·74	0·037	0·08	870	365	0·6–1·0	265·5
410 (annealed)	770	609	10	207	7·7	11	0·058	0·08	1 050	830	0·46	94
nitronic 33 60% (cold worked)	1 428	1 281	6·5	189	7·9	15·0	0·04	0·08	2 261	1 764	0·8	131

Titanium alloys												
Ti–5Al–2·5Sn	770	700	14	112	4·46	9·36	0·016	0·09	1 295	1 225	2·05	176·4
Ti–6Al–4V	980	875	12	112	4·43	9·4	0·016	0·09	1 540	1 435	2·25	178·5
Superalloys												
Inconcel X–750	1 260	840	30	217	8·51	11·5	0·33	0·07	1 505	945	1·87–2·13	368
Inconel 718	1 365	1 190	15	217	8·51	11·5	0·31	0·07	1 680	1 505	1·87–2·13	239
Polymers												
Teflon	21	14	2·0	0·42	2·16	99	0·001	0·132	105	95	30	2
Epoxy glass fibre laminates	525	300	1·5	50	2·5	80	0·0010	0·14	700	500	70	9
Copper alloys												
OFHC	210	60	60	120	8·94	16·8	0·81	0·07	364	90	0·9–0·97	136
70Cu–30Zn (annealed)	350	200	75	112	8·53	19·9	0·29	0·06	518	210	0·8–0·9	273
70Cu–30Zn (hard)	532	441	10	112	8·53	19·9	0·29	0·06	790	465	0·8–0·9	63
70–30 Cupro-nickel (annealed)	385	126	36	154	8·94	16·2	0·07	0·07	560	210	1·3–1·5	139

*R T = room temperature

**Thermal conductivity is given in cal/cm^2/cm/°C/s, and specific heat is given in cal/g/°C.

†For definition of toughness index, see text.

pair their corrosion resistance. Martensitic transformation in austenitic stainless steel can occur on cooling to 77 K, which makes the material very brittle. Generally high percentages of Ni, Mn and C tend to stabilise the austenite, while Cr and other BCC metals have the opposite effect.

Titanium alloys have very high strength/weight ratios and higher toughness than aluminium alloys but are also more expensive.

The low ductility and toughness of polymers at low temperatures is a serious disadvantage in their use for storage vessel applications where impact loads may be encountered and some fibre reinforcement is necessary. The main advantage of these composite materials is their high strength/weight ratio but their properties are highly directional and they may have porosity if produced by filament winding. High cost is another disadvantage of fibre reinforced polymers.

19.5 ALLOCATION OF WEIGHTING FACTORS

In the case of low temperature applications, toughness at the service temperature is the main selection criterion.

The room temperature yield strength and elastic modulus of the material are also important in determining the thickness and rigidity of the vessel. With stronger material, thinner walls can be used which means a lighter vessel and less cool-down losses. The lower the specific gravity the lighter will be the vessel; and the lower the thermal conductivity the lower will be the heat losses. Lower specific heat is also desirable to reduce cool-down losses, and a lower thermal expansion coefficient reduces thermal stresses. Cost is also included in the comparison of the materials as the vessel has to be produced at the lowest possible cost. The relative importance of the different parameters is shown in Table 19.3. The toughness is most

TABLE 19.3

WEIGHTING FACTORS FOR LOW TEMPERATURE MATERIAL PROPERTIES

Property	*Weighting factor*
Toughness	0·22
Room temperature yield strength	0·15
Room temperature Young's modulus	0·11
Specific gravity	0·11
Thermal expansion coefficient	0·11
Thermal conductivity	0·11
Specific heat	0·08
Cost	0·11

important, followed by room temperature yield strength. The other parameters are equal in importance except for specific heat which is considered least important. This is because cool-down losses can be reduced by purging with a cheap cryogen to cool the vessel first. Also it is a transient stage in the storage, and losses from it are expected to be a small fraction of the total losses of storage.

19.6 DECIDING ON THE OPTIMUM MATERIAL

Decision on the optimum material was reached using the weighted properties method described in Chapter 13, and the results are given in Table 19.4. The full hard 301 stainless steel is the obvious first choice. Its

TABLE 19.4

EVALUATION AND GRADING OF LOW TEMPERATURE MATERIALS

Material	*Performance index* γ	*Preference*
Aluminium alloys		
2014–T6	37·6	11
5052–0	34·3	16
5083–0	34·9	14
5456–0	34·8	15
7075–76	36·2	13
Stainless steels		
301 (full hard)	70	1
304 (annealed)	41	10
310 (75% cold worked)	52·8	2
321 (annealed)	41·7	9
410 (annealed)	49·2	6
Nitronic 33 (60% cold worked)	49·7	5
Titanium alloys		
Ti–5 Al–2·5 Sn	47	7
Ti–6 Al–4 V	45·2	8
Superalloys		
Inconel X-750	51·1	4
Inconel 718	52·0	3
Polymers		
Teflon	29	20
Epoxy glass fibre laminates	31·4	19
Copper alloys		
O.F.H.C.	26·3	21
70Cu–30Zn (annealed)	37·4	12
70Cu–30Zn (hard)	34·0	17
70–30 Cupronickel (annealed)	32·7	18

excellent toughness, strength and modulus of elasticity are the main assets. The cost of this material is also reasonable. Three materials follow in the order of merit, 75% cold worked 310 stainless steel, Inconel 718 and Inconel X-750. These materials combine reasonable mechanical and thermal properties as shown in Table 19.2.

Although aluminium alloys are the cheapest on the list, their poor mechanical properties and relatively high specific heat and conductivity are serious drawbacks.

BIBLIOGRAPHY

1. G.G. Haselden, *Cryogenic Fundamentals*, Academic Press (London), 1971.
2. *Materials Selector, Materials Engineering*, 1971.
3. *Metal Progress, Data Book*, American Soc. for Metals, 1977.
4. *Metals Handbook*, Vol. 1, 'Properties and Selection of Metals', 8th Edn, American Soc. for Metals, 1961.
5. R.B. Scott, *Cryogenic Engineering*, D. Van Nostrand (London), 1959.
6. R.W. Vance and W.M. Duke, Eds, *Applied Cryogenic Engineering*, John Wiley and Sons (New York), 1962.

20

Selection of Materials for Protective Coating

20.1 INTRODUCTION

Metallic materials used in construction of engineering and consumer products are usually not adequately resistant to the environments in which they will be used, and this makes them subject to corrosion and deterioration. Also, the metallic surfaces produced by the common production techniques do not possess inherently attractive colours and thus are lacking in sales appeal. As a consequence of these conditions, most manufactured products, after achieving the desired shape, require one or more finishing operations. The cost of the finish can be an appreciable percentage of total factory cost. Finishes can be either an additive coating or a conversion layer. In additive coatings a layer of material is applied to the surface to be finished. Common examples are paints, lacquers, varnishes and plated coatings. In conversion finishes, the surface to be finished is changed by the application of either a reactive substance, or by thermal, electrochemical or mechanical action. Examples are anodising, sheradising and diffusion coatings.

Good adhesion is an important requirement for a successful coating. Clean surfaces and controlled temperature and humidity are necessary for adhesion. The main methods of removing dirt and oils are water base solvents, chemical solvents and ultrasonic vibrations; acid pickling is used for removing oxides. The main methods of drying, temperature control and humidity control are infra-red heating, steam and oil or gas fire.

20.2 USES AND FUNCTIONS OF COATINGS

Coatings are usually required to perform one or more of the following functions:

1. Protection against corrosion or oxidation.
2. Protection against abrasion and wear.

3. Provision of electrical and thermal conductivity or insulation.
4. Control of surface appearance, colour or polish.

Regardless of its function, a coating must adhere to the base material. Both the surface texture and cleanliness can affect the degree of adhesion. Rougher, cleaner surfaces enhance adhesion. The appearance of coatings can be measured by their colour, brightness, reflectivity and opacity. The colour retention indicates the length of time during which the coating will not fade and will remain attractive under normal service conditions.

Industrially, protection characteristics of a coating are of prime importance. Protection can be achieved in two ways: 1, isolation of the surface from the environment; 2, electrochemical action. The former function is usually performed by non-metallic coatings and thickness, soundness and strength of the coating will control its effectiveness as an isolator. The second function is achieved when the coating is anodic with respect to the base metal, so that it will dissolve anodically, while the base material, which is the cathode in the galvanic cell, will not be attacked. This latter function is usually performed by coatings that include metallic materials. Examples are zinc, cadmium and aluminium which are above iron in the galvanic series and will function as anodic coatings on steel. The two ways of protection can be combined by first applying an anodic coating to the base metal and then a non-metallic finish. Organic materials are the most widely used finishing coating not only for metals but also for other construction engineering materials. It has been estimated that over 50% of all metal surfaces for which an impervious, pore-free, adherent and attractive property is required are treated with organic paints. As a family, organic finishes are relatively inexpensive, provide the opportunity for a variety of pleasing colours and give good resistance to corrosion. However, they do not allow the holding of close dimensional tolerances and they have only average resistance to abrasion. Their resistance to elevated temperatures is less than that of metallic or ceramic coatings. These points of strength and weakness should be borne in mind when employing organic finishes. The following discussion will be devoted mainly to the selection of the organic finishing coatings.

20.3 ORGANIC COATINGS

Organic coatings are mostly additive type finishes and can be applied in one or more layers with a thickness that varies from 5 to 500 microns. Organic coatings depend mainly on their chemical inertness and impermeability in providing protection against corrosion. Organic coatings may be applied by brush, knife, dip, roller, flow, tumble, silk screen or spray methods. Brushing is easy but slow and usually other methods are used for production. Dipping can be mechanised easily but requires paints that film out and workpieces that can be immersed easily and contain no pockets. Flow coating is done by pouring paint onto workpieces and recirculating the runoff; however, spraying is the most widely used method of industrial painting as it is fast, dependable, versatile and uniform.

An organic coating is made up of two principle components: a vehicle and a pigment. The vehicle contains the film forming ingredients that dry to form the solid film. It also acts as a carrier for the pigment. Vehicles can be either oil or resin, but oils have limited use in industrial finishes. Nearly all polymers can be used as film formers and frequently two or more kinds are combined to give the set of properties desired. The properties of polymers as coatings are usually similar to those of the bulk form. Pigments, which may or may not be present, give the required colour, opacity and flow characteristics, and can also contribute to the protection against corrosion of the base metal and against the destructive action of ultra-violet light rays on the polymeric vehicle.

Varnishes consist of thermosetting resins and oils. They are clear and unpigmented and can be used alone for coating but their major use in industrial finishing is as a vehicle. Enamels consist of intimate dispersions of pigments in a varnish or a resin vehicle or in a combination of both. They may dry by oxidation at room temperature and/or by polymerisation at room or elevated temperatures. Enamels as a class are hard and tough and resist attack by most chemical agents. Because of their wide range of useful properties, enamels are the most widely used organic coatings in industrial applications. Among the areas where enamels are extensively used are household appliances, automotive products and railway equipment.

Lacquers are noted primarily for their fast drying as they depend only on solvent evaporation, but at least two coats of lacquer are required to give the protection afforded by one coat of varnish or enamel. Their durability for exterior application and adhesion to metallic surfaces are poor. However, their fast drying sometimes outweighs their drawbacks, e.g. in high speed production.

20.4 BEHAVIOUR OF ORGANIC COATINGS

While for indoor decorative purposes, a finish that gives the necessary adhesion, abrasion resistance and colour retention is considered satisfactory, industrial, and outdoor applications usually require additional qualities, e.g. durability under exterior conditions of direct sunlight, moisture and temperature changes; resistance to corrosive atmosphere; resistance to certain chemicals; and resistance to high temperature without deterioration.

Of the many reasons for failure of organic finishes, the following are the most common:

1. Blistering, due to poor adhesion or due to wide temperature changes.
2. Mechanical damage, e.g. chipping of brittle coatings and scratching of soft ones.
3. Softening or discolouring due to contact with destructive agents, e.g. corrosive atmospheres or chemicals.
4. Discolouration due to the action of light or slow oxidation of the vehicle.

No one organic finish can meet all these requirements. Nor will an organic finish best suited to withstand a particular destructive agent be expected to last indefinitely. However, selection of the most appropriate coating for a given application will give better finish and longer life at a lower cost. To arrive at the best compromise available, it is important to specify the service conditions and to decide which property to sacrifice for the sake of another.

Besides withstanding the service conditions, a finish should be achieved at the lowest possible cost. Material costs for finishing are usually based on the cost of material to cover a unit area at a given thickness when dry. This is a more accurate basis for comparison than cost per unit weight or per unit volume because finishing materials vary in density and coverage. The cost will also depend on the method of application of the finish, e.g. dipping can be about 35% more efficient than spraying due to the absence of overspray. The total finishing cost includes the cost of cleaning and labour, but these vary little between the different materials and may be omitted in comparisons between different coatings.

20.5 ALLOCATION OF WEIGHTING FACTORS

The above discussion shows that the main material requirements for protective coatings can be summarised as follows:

1. Abrasion resistance.
2. Flexibility.
3. Adhesion.

4. External durability.
5. Colour retention.
6. Resistance to atmosphere.
7. Resistance to chemicals.
8. Maximum service temperature.
9. Cost.

The first two properties correspond to the mechanical properties of the dry coating. Abrasion resistance represents the resistance to scratching and loss of gloss due to rubbing of foreign bodies on the coating, and flexibility represents the resistance to cracking due to small changes in configuration or dimensions of the base surface. Stronger adhesion reduces the possibility of the coating blistering and peeling off, and better external durability indicates slower deterioration due to sunlight, temperature changes and mositure. Resistance to fading due to oxidation of vehicle and the action of light is represented by colour retention, and resistance to destructive agents is represented by properties 6 and 7. In applications where the coated surface may be heated to high temperatures, the maximum service temperature of the coating material should be higher than the expected surface temperature in order to avoid permature discolouration or even destruction of coating. As in all engineering application, satisfactory coating should be achieved at the least possible cost on the job. Table 20.1 gives a list of names and properties of the commonly used organic coating materials. The different properties of the commonly in the form of relative ratings with 3 representing *Excellent*; 2, *Very good–good*; 1, *Fair*; and 0, *Poor*. In the case of cost the cheapest material is given 3, the medium is given 2, the expensive is given 1, and the very expensive is given 0.5.

For the present case study, selection will be made for coating materials for motor car bodies. The relative importance of the 9 requirements discussed above was determined using the digital logic approach, described in Chapter 8, and Table 20.2 gives the weighting factors.

The weighted properties method, as outlined in Chapter 13, was used to obtain the figures of merit of Table 20.3. The rating of materials with respect to their performance index (γ) is also shown in Table 20.3.

20.6 DECIDING ON THE OPTIMUM MATERIAL

For the present case study, materials with a rating of poor (0) in any of the properties were eliminated as unsuitable. The material with the highest performance index (γ) was found to be acrylic, which combines excellent

TABLE 20.1

RATING OF ORGANIC COATINGS

	Cost	*Abrasion resistance*	*Flexibility*	*Adhesion*	*Resistance to atmosphere (salt spray)*	*Exterior durability*	*Colour retention*	*Resistance to chemicals (general)*	*Maximum service temperaature rating*
Alkyd	3	2	3	3	1	3	1	1	1
Amine-alkyd	3	3	2	3	1	3	1	1	2
Acrylic	2	2	3	2	3	3	3	1	1
Cellulose (Butyrate)	1	2	3	2	3	2	3	1	1
Epoxy	1	3	3	3	3	1	1	3	2
Epoxy ester	2	3	1	3	3	2	1	1	2
Fluorocarbon	0·5	1	1	2	3	3	1	3	2
Phenolic	2	3	1	3	3	3	0	2	2
Polyamide	2	3	1	2	1	0	2	1	2
Plastisol	3	3	3	2	3	2	1	3	1
Polyester (oil free)	2	2	2	3	3	2	2	1	1
Polyvinyl fluoride (PVF)	0·5	3	3	2	3	3	2	3	1
Polyvinylidene fluoride (PVF2)	0·5	3	3	2	3	3	2	3	1
Silicone	1	2	1	1	3	3	3	1	3
Silicone alkyd	1	2	1	2	2	3	2	2	3
Silicone polyester	1	2	2	2	3	3	2	2	3
Silicone acrylic	1	2	1	2	2	3	3	2	3
Vinyl	2	2	3	1	3	3	2	1	1
Vinyl alkyd	2	2	2	2	2	1	2	1	1
Polyvinyl chloride (PVC)	1	3	3	3	3	2	1	3	1
Neoprene (rubber)	3	3	3	2	3	3	1	1	1
Urethane	0·5	3	3	3	3	3	1	1	2

TABLE 20.2

WEIGHTING FACTORS FOR PROTECTIVE COATINGS FOR MOTOR CAR BODIES

Property	*Weighting factor*
Cost	0·10
Abrasion resistance	0·10
Flexibility	0·07
Adhesion	0·20
Resistance to atmosphere (salt spray)	0·15
Exterior durability	0·20
Colour retention	0·15
Resistance to chemicals (general)	0·03
Maximum service temperature	0·00

TABLE 20.3

EVALUATION AND GRADING FOR PROTECTIVE COATINGS FOR MOTOR CAR BODIES

Material	*Performance index* (γ)	*Preference*
Alkyd	2·24	9
Amine-alkyd	2·27	7
Acrylic	2·54	1
Cellulose (Butyrate)	2·24	9
Epoxy	2·1	13
Epoxy ester	2·2	10
Fluorocarbon	1·91	15
Phenolic	—	rejected
Polyamide	—	rejected
Plastisol	2·3	6
Polyester (oil free)	2·32	5
Polyvinylfluoride (PVF)	2·4	3
Polyvinylidene fluoride (PVF2)	2·4	3
Silicone	2·1	13
Silicone alkyd	2·03	14
Silicone polyester	2·25	8
Silicone acrylic	2·18	12
Vinyl	2·19	11
Vinyl alkyd	1·77	16
Polyvinyl chloride (PVC)	2·3	6
Neoprene	2·44	2
Urethane	2·39	4

flexibility, resistance to atmosphere, durability under exterior conditions and colour retention. Its abrasion resistance is very good and cost is reasonable. The second best material is neoprene, followed by polyvinylfluoride (PVF) and polyvinylidene fluoride (PVF2). Urethane and polyester come fourth and fifth respectively.

BIBLIOGRAPHY

1. H.R. Clauser, *Industrial and Engineering Materials,* McGraw-Hill (New York), 1975.
2. E.P. De Garmo, *Materials and Processes in Manufacturing,* 4th Edn, Macmillan (New York), 1974.
3. D.R. Gabe, *Principles of Metal Surface Treatment and Protection,* Pergamon (Oxford, New York), 1972.
4. *Metal Progress, Data Book,* American Soc. for Metals, 1977.
5. B.W. Niebel, and A.B. Draper, *Product Design and Process Engineering,* McGraw-Hill (New York), 1974.
6. J.F. Young, *Materials and Processes,* 2nd Edn, John Wiley and Sons (New York), 1954.

Appendix A.1

Conversions to SI units

Quantity	*Multiply number of:*	*by*	*to obtain number of:*
Length	Inches	25·4	mm
	Feet	0·304 8	metres (m)
	Yards	0·914 4	metres (m)
Area	Square inches	645·16	mm^2
	Square feet	0·092 903	m^2
	Square yards	0·836 130	m^2
Volume	in^3	1 6387·1	mm^3
	ft^3	0·028 316 8	m^3
	yd^3	0·764 555	m^3
Mass	Ounces	0·028 349 5	kilogrammes (kg)
	Pounds	0·453 592 37	kg
	Short tons	907·185	kg
	Long tons	1 016·05	kg
Density	lb/in^3	27 679·9	kg/m^3
	lb/ft^3	16·018 5	kg/m^3
Force	Pounds force (lbf)	4·448 22	newton (N)
	Tons force (long)	9 964·02	N
	Dynes	10^{-5}	N
	kgf	9·806 65	N
Stress	lbf/in^2	$6{\cdot}894\ 76 \times 10^3$	N/m^2
	$tonf/in^2$	$15{\cdot}444\ 3 \times 10^6$	N/m^2
	kgf/cm^2	$98{\cdot}066\ 5 \times 10^3$	N/m^2
Work	ft/lbf	1·355 82	Joule (J)
	hp/h	$2{\cdot}684\ 52 \times 10^6$	J
	BTU	$1{\cdot}055\ 06 \times 10^3$	J
	kw/h	$3{\cdot}6 \times 10^6$	J
	kcal	$4{\cdot}186\ 8 \times 10^3$	J
	kgf/m	9·806 65	J
Power	ft/lbf s	1·355 82	Watt (W)
	Horsepower	745·7	W
	Metric hp (CV)	735·499	W
	BTU/h	0·293 071	W
Thermal conduc-tivity	BTU/hr ft °F	1·730 73	Wm^{-3}
	BTU in/(hr ft ^{2} °F)	0·144 228	$Wm^{-1}K^{-1}$
	kcal/(mh°C)	1·163	$Wm^{-1}K^{-1}$

Temperature °C = 5/9)°F−32) = 5/9(°R−459·69) = K−273·15
K = 5/9(°R) = 5/9(°F−32) + 273·15 = °C + 273·15
= 5/9(°F) × 255·37

Appendix A.2

Temperature Conversions*

°F	*Temperature to be converted*	°C	K
−418	−250	−156·7	116·5
−400	−240	−151·1	122·1
−382	−230	−145·6	127·6
−364	−220	−140·0	133·2
−346	−210	−134·4	138·8
−328	−200	−128·9	144·3
−310	−190	−123·3	149·9
−292	−180	−117·8	155·5
−274	−170	−112·2	161·1
−256	−160	−106·7	166·5
−238	−150	−101·1	172·1
−220	−140	−95·6	177·6
−202	−130	−90·0	183·2
−184	−120	−84·4	188·8
−166	−110	−78·9	194·3
−148	−100	−73·3	199·9
−130	−90	−67·8	205·4
−112	−80	−62·2	211·0
−94	−70	−56·7	216·5
−76	−60	−51·1	222·1
−58	−50	−45·6	227·6
−40	−40	−40·0	233·2
−22	−30	−34·4	238·8
−4	−20	−28·9	244·3
+14	−10	−23·3	249·9
+32	±0	−17·8	255·4
+50	+10	−12·2	261·0
+68	+20	−6·7	266·5
+86	+30	−1·1	272·1
+104	+40	+4·4	277·6
122	50	10·0	283·2

(*Contd.*)

APPENDIX A. 2 (*Contd.*)

°*F*	*Temperature to be converted*	°*C*	*K*
140	60	15·6	288·8
158	70	21·1	294·3
176	80	26·7	299·9
194	90	32·2	305·4
212	100	37·8	311·0
230	110	43·3	316·5
248	120	48·9	322·1
266	130	54·4	327·6
284	140	60·0	333·2
302	150	65·6	338·8
320	160	71·1	344·3
338	170	76·7	349·9
356	180	82·2	355·4
374	190	87·8	361·0
392	200	93·3	366·5
410	210	98·9	372·1
428	220	104·4	377·6
446	230	110·0	382·2
464	240	115·56	388·7
482	250	121·1	394·3
500	260	126·7	399·9
518	270	132·2	405·4
536	280	137·8	411·0
554	290	143·3	416·5
572	300	148·9	422·1
590	310	154·4	427·6
608	320	160·0	433·2
626	330	165·6	438·8
644	340	171·1	444·3
662	350	176·7	449·9
680	360	182·2	455·4
698	370	187·8	461·0
716	380	193·3	466·5
734	390	198·9	472·1
752	400	204·4	477·6
770	410	210·0	483·2
788	420	215·6	488·8
806	430	221·1	494·3
824	440	226·7	499·9

(*Contd.*)

APPENDIX A. 2 (*Contd.*)

°*F*	*Temperature to be converted*	°*C*	*K*
842	450	232·2	505·4
860	460	237·8	511·0
878	470	243·3	516·5
896	480	248·9	522·1
914	490	254·4	527·6
932	500	260·0	533.2
950	510	265·6	538·8
968	520	271·1	544·3
986	530	276·7	549·9
1004	540	282·2	555·4
1022	550	287·8	561·0
1112	600	315·6	588·8
1202	650	343·3	616·5
1242	700	371·1	644·3
1382	750	398·9	672·1
1472	800	426·7	699·9
1562	850	454·4	727·6
1652	900	482·2	755·4
1742	950	510·0	783·2
1832	1000	537·8	811·0
1922	1050	565·6	838·8
2012	1100	593·3	866·5
2102	1150	621·1	894·3
2192	1200	648·9	922·1
2282	1250	676·7	949·9
2372	1300	704·4	977·6
2462	1350	732·2	1005·4
2552	1400	760·0	1033·2
2642	1450	787·8	1061·0
2732	1500	815·6	1088·8
2822	1550	843·3	1116·5
2912	1600	871·1	1144·3
3002	1650	898·9	1172·1
3092	1700	926·7	1199·9
3182	1750	954·4	1227·6
3272	1800	982·2	1255·4
3362	1850	1010·0	1283·2
3452	1900	1037·8	1311·0

(*Contd.*)

APPENDIX A. 2 (*Contd.*)

°F	*Temperature to be converted*	°C	K
3542	1950	1065·6	1338·8
3632	2000	1093·3	1366·5
3722	2050	1121·1	1394·3
3812	2100	1148·9	1422·1
3902	2150	1176·7	1449·9
3992	2200	1204·4	1477·6
4082	2250	1232·2	1505·4
4172	2300	1260·0	1533·2
4262	2350	1287·8	1561·0
4352	2400	1315·6	1588·8
4442	2450	1343·3	1616·5
4532	2500	1371·1	1644·3
4622	2550	1398·9	1672·1
4712	2600	1426·7	1699·9
4802	2650	1454·4	1727·6
4892	2700	1482·2	1755·5
4982	2750	1510·0	1783·2
5072	2800	1537·8	1811·0

*If the temperature to be converted is in °F, read °C and K in the right-hand column. If in °C, read °F in the left-hand column.

Appendix A.3

Imperial/SI Stress Conversion Factors*

ksi	*Stress to be converted*	*MN/m²*
—	0	—
0·15	1	6·89
0·29	2	13·79
0·44	3	20·68
0·58	4	27·57
0·73	5	34·47
0·87	6	41·37
1·02	7	48·26
1·16	8	55·16
1·31	9	62·05
1·45	10	68·95
1·60	11	75·84
1·74	12	82·74
1·89	13	89·63
2·03	14	96·53
2·18	15	103·4
2·32	16	110·3
2·47	17	117·2
2·61	18	124·1
2·76	19	131·0
2·90	20	137·9
3·05	21	144·8
3·19	22	151·7
3·34	23	158·6
3·48	24	165·5
3·63	25	172·4
3·77	26	179·3
3·92	27	186·2
4·06	28	193·1
4·21	29	199·9
4·35	30	206·8
4·50	31	213·7
4·64	32	220·6
4·79	33	227·5

(*Contd.*)

APPENDIX A.3 (*Contd.*)

ksi	*Stress to be converted*	MN/m^2
4·93	34	234·4
5·08	35	241·3
5·22	36	248·2
5·37	37	255·1
5·51	38	262·0
5·66	39	268·9
5·80	40	275·8
5·95	41	282·7
6·09	42	289·6
6·24	43	296·5
6·38	44	303·4
6·53	45	310·3
6·67	46	317·2
6·82	47	324·1
6·96	48	331·0
7·11	49	337·8
7·25	50	344·7
7·40	51	351·6
7·54	52	358·5
7·69	53	365·4
7·83	54	372·3
7·98	55	379·2
8·12	56	386·1
8·27	57	393·0
8·41	58	399·9
8·56	59	406·8
8·70	60	413·7
8·85	61	420·6
8·99	62	427·5
9·14	63	434·4
9·28	64	441·3
9·43	65	448·2
9·57	66	455·1
9·72	67	462·0
9·86	68	468·8
10·01	69	475·7
10·15	70	482·6
10·30	71	489·5
10·44	72	496·4
10·59	73	503·3

(*Contd.*)

APPENDIX A.3 (*Contd.*)

ksi	*Stress to be converted*	*MN/m²*
10·73	74	510·2
10·88	75	517·1
11·02	76	524·0
11·17	77	530·9
11·31	78	537·8
11·46	79	544·7
11·60	80	551·6
11·75	81	558·5
11·89	82	565·4
12·04	83	572·3
12·18	84	579·2
12·33	85	586·1
12·47	86	593·0
12·62	87	599·8
12·76	88	606·7
12·91	89	613·6
13·05	90	620·5
13·20	91	627·4
13·34	92	634·3
13·49	93	641·2
13·63	94	648·1
13·78	95	655·0
13·92	96	661·9
14·07	97	668·8
14·21	98	675·7
14·36	99	682·6
14·50	100	689·5

*If the stress to be converted is in ksi (10^3 psi), read MN/m^2 in the right hand column. If in MN/m^2 read ksi in the left-hand column. 1 $MN/m^2 = 0{\cdot}145\ 0377$ ksi, and 1 ksi $= 6{\cdot}894\ 759\ MN/m^2$

Appendix A.4

Hardness Conversions
Hardened Steel and Hard Alloys

Rockwell scale			VHN	BHN	Tensile strength
C	A	30N	10 kg	3000 kg	MN/m^2
80	92·0	92·0	1 865	—	—
75	89·5	89·0	1 478	—	—
70	86·5	86·0	1 076	—	—
65	84·0	82·0	820	—	—
64	83·5	81·0	789	—	—
62	82·5	79·0	739	—	—
60	81·0	77·5	695	614	2 310
58	80·0	75·5	655	587	2 205
56	79·0	74·0	617	560	2 065
54	78·0	72·0	580	534	2 006
52	77·0	70·5	545	509	1 889
50	76·0	68·5	513	484	1 758
48	74·5	66·5	485	460	1 634
46	73·5	65·0	458	437	1 524
44	72·5	63·0	435	415	1 427
42	71·5	61·5	413	393	1 338
40	70·5	59·5	393	372	1 255
38	69·5	57·5	373	352	1 179
36	68·5	56·0	353	332	1 117
34	67·5	54·0	334	313	1 054
32	66·5	52·0	317	297	9 929
30	65·5	50·5	301	283	937·7
28	64·5	48·5	285	270	889·4
26	63·5	47·0	271	260	848
24	62·5	45·0	257	250	807
22	61·5	43·0	246	240	772
20	60·5	41·5	236	230	745

Appendix A.5

Hardness Conversions
Soft Steel, Grey and Malleable Cast Iron, and Most non-Ferrous Metals

Rockwell scale			*BHN* *500 kg*	*VHN 10 kg* *and*	*Tensile* *strength*
B	*A*	*30T*	(10mm ball)	*BHN 3000kg*	MN/m^2
100	61·5	82·0	201	240	800
98	60·0	81·0	189	228	752
96	59·0	80·0	179	216	710
94	57·5	78·5	171	205	676
92	56·5	77·5	163	195	641
90	55·5	76·0	157	185	614
88	54·0	75·0	151	176	586
86	53·0	74·0	145	169	559
84	52·0	73·0	140	162	538
82	50·5	71·5	135	156	517
80	49·5	70·0	130	150	497
78	48·5	69·0	126	144	476
74	46·0	66·0	118	135	448
70	44·0	63·5	110	125	420
66	42·0	60·5	104	117	392
62	40·5	58·0	98	110	—
58	38·5	55·0	92	104	—
54	37·0	52·5	87	98	—
50	35·0	49·5	83	93	—
46	33·5	47·0	79·5	88	—
42	31·5	44·0	76·0	86	—
38	30·0	41·5	73·0	—	—
34	28·0	38·5	70·0	—	—
30	26·5	36·0	67·0	—	—
20	22·0	29·0	61·5	—	—
10	—	22·0	57·0	—	—

Appendix A.6

Imperial*SI Impact Strength Conversion Factors*

Joule (J)	*Impact strength to be converted*	*ft/lb*
0·14	0·1	0·07
0·27	0·2	0·15
0·41	0·3	0·22
0·54	0·4	0·30
0·68	0·5	0·37
0·81	0·6	0·44
0·95	0·7	0·52
1·08	0·8	0·59
1·22	0·9	0·66
1·36	1·0	0·74
2·71	2·0	1·48
4·07	3·0	2·21
5·42	4·0	2·95
6·78	5·0	3·69
8·13	6·0	4·43
9·49	7·0	5·16
10·8	8·0	5·90
12·2	9·0	6·64
13·6	10.0	7·38
14·9	11	8·11
16·3	12	8·85
17·6	13	9·59
19·0	14	10·3
20·3	15	11·1
21·7	16	11·8
23·0	17	12·5
24·4	18	13·3
25·8	19	14·0
27·1	20	14·8
40·7	30	22·1
54·2	40	29·5

(*Contd.*)

APPENDIX A.6 (*Contd.*)

Joule (J)	*Impact strength to be converted*	*ft/lb*
67·8	50	36·9
81·3	60	44·3
94·9	70	51·6
108·4	80	59·0
122·0	90	66·4
135·6	100	73·8

*If the impact strength to be converted is in ft/lbf, read Joules (J) in the left hand column. If in joules (J), read ft/lbf in the right hand column. 1 ft/lbf = 1·355 818 joule and 1 joule = 0·737 562 ft/lbf.

Appendix A.7

Metric Equivalents of Fractional Inches

1/4	*1/8*	*1/16*	*1/32*	*MM*
			1/32	0·79
		1/16		1·59
			3/32	2·38
	1/8			3·18
			5/32	3·97
		3/16		4·76
			7/32	5·56
1/4				6·35
			9/32	7·14
		5/16		7·97
			11/32	8·75
	3/8			9·53
			13/32	10·32
		7/16		11·11
			15/32	11·91
1/2				12·70
			17/32	13·49
		9/16		14·29
			19/32	15·08
	5/8			15·88
			21/32	16·67
		11/16		17·46
			23/32	18·26
3/4				19·05
			25/32	19·84
		13/16		20·64
			27/32	21·43
	7/8			22·23
			29/32	23·02
		15/16		23·81
			31/32	24·61
1				25·40

Index